PROFESSIONAL
ELECTRONIC
BEST PRACT

PROFESSIONAL ELECTRONIC DESIGN BEST PRACTICES

Matthew F. Berger

Professional Electronic Design Best Practices
Copyright © 2024 Matthew F. Berger

All Rights Reserved. No part of this book may be reproduced, scanned, or transmitted in any form, digital, audio or printed, without the expressed consent of the author.

S/P Publishing, Winterthur, Switzerland

ISBN: 979-8390127971

Contents

Foreword   9

Using this Book   2

1   Component Selection   3

     Connector Selection   3
     Pushbutton and Switch Selection   5
     Resistor Selection   7
     Potentiometer Selection   9
     Trimmer Selection   9
     Capacitor Selection   10
     Selecting Power Inductors, RF Coils and Chokes   15
     Additional Passive Component Considerations   17
     Semiconductor Selection   20
     Diodes and LEDs   20
     Transistors   21
     Optocouplers   21
     Operational Amplifiers   21
     Reference Voltage Sources   23
     Choosing Integrated Circuits for Simple Logic Tasks   26
     Memory Selection   26
     Analog-to-Digital Converter Selection   28
     Digital-to-Analog Converter Selection   28
     Microcontroller Selection   31
     More than one Microcontroller?   33
     Safety Microcontrollers   34
     Microcontroller GPIO Pin Notes   35
     Microcontroller Clocking Notes   36
     Microcontroller External Memory Notes   36
     Microcontroller Reset Notes   37
     Why You Should Use an External Watchdog   38
     Notes on Low Level Programming and HAL   39
     Notes on Debugging Microcontroller Firmware   40
     Programming the Microcontroller in Production   41
     Choosing Field-Programmable Gate Arrays   41

vi Contents

Notes on FPGA Reset and FPGA Clocking	43
System-on-Chip Selection	44
Integrated Circuit Package Considerations	45
Component Selection and Obsolescence	47
Potential Problems with the Assembler	51
Sensor Selection	51
Display Selection	52
Selecting Off-the-Shelf Modules	53
Using Wireless Modules	54
Do Not Program Apps Yourself	56
Remote Firmware Update is Complicated	56
BOM Consolidation	57
Visits by Component Vendors	57

2 Direct-Current Supply Design 59

Power Supply	59
Fuses	61
Reverse Polarity Protection Ideas	62
Dealing with Inrush Currents	64
Battery and Accumulator Selection	65
Batteries and Voltage Regulators	66
Batteries and Charge Pumps	68
Boost Converter for Single Cell Operation	69
Batteries and a Step-Down Converter	69
Batteries and Buck-Boost Converters	70
DC/DC Converter Selection	70
DC/DC Converter Input/Output Filtering	71
Pi Filters: Notes on Chokes and Ferrite Beads	72
Pi Filters: Notes on Capacitors	73
Pi Filter Variants	75
DC/DC Converter Layout	76
Low-Noise Switchers	79
Cascaded, Series, and Parallel DC/DC Converters	79
DC/DC Converter Limitations	81
Current Measurement Variants	83
Maximum Currents in Traces and Vias	85
Maximum DC Source Impedance	86
Power Supply Decoupling Without a Power Plane	87
Power Supply Decoupling with a Power Plane	92
Decoupling Capacitors and the Time Domain	96
Some Additional Notes about Decoupling Capacitors	97
Solutions When Decoupling Caps Are Not Enough	99
Chokes in the Supply Path	99
Multiple Ground Planes, or one Large Ground Plane?	100
Ground Plane to Chassis Connection	103
Multiple Power Planes	104

3	Robust Interfaces	105	

Notes on the I^{2}C Interface	105
Notes on the SPI Interface	106
Low Voltage Differential Signaling Explained	107
RS-422, RS-485, CAN, and USB	109
Electrostatic Discharge Protection	109
ESD Protection for High-Speed Interfaces	115
ESD Protection for Switches and Keypads	116
ESD Protection for Instrumentation Amplifiers	116
ESD Protection Using a Capacitor	118
ESD Protection Using a Varistor	118
Plastic Enclosures and ESD	118
Burst Protection	119
Surge Protection	120
Burst and Surge Protection Example	122
Overvoltages due to Long Cables	122
Earth Potential Discrepancy and Differential Signals	123
Interference Protection Measures	124
Choosing the Right Cable Length	126
Chokes in the Signal Path?	127
Using Three-Terminal Capacitors	128
Using Common-Mode Chokes	129
Cable Twisting and Differential Transmission	130
Cable Shielding	131
Cable Strands and Cable Routing	132
Interface Example Designs	135
Analog Differential Transmission?	138
Some Advice on EMC Testing	140

4	Signals on Printed Circuit Boards	141	

Enclosure Shielding: Ideal and Real Faraday Cage	141
How to Manage Openings	143
Ground Plane Advantages and Problems	143
How to Avoid Crosstalk	145
Flooding Signal Planes	148
Mixed-Signal Design Aspects	148
Improving the Quality of Digital Signal Connections	151
How to Avoid Overshoot	154
Trace Length Limits	158
Signal Trace Impedance and Termination	161
Branching at Fast Digital Signals, Parallel Termination	165
Unidirectional Signal Series Termination	169
Termination Bidirectional Connections	169
Terminating a Bidirectional Bus	170
Differential Signal Routing Details	171
Tracks at Right Angles	173

viii Contents

Vias and Fast Signals	173
Tri-State Considerations	174

5 Thermal Management 175

Caution Defining the Ambient Temperature Range	176
Common and Unusual Suspects That Get Hot	177
Estimate of the Internal Case Temperature	177
Placing Hot Components in a Closed Case	179
Placing a Heat Sink Outside the Case	179
Maximum Contact Temperature	180
Questionable Published Heatsink Resistance Values	180
Heat Sink Assembly in Production	181
Heat Sink and Electrical Insulation	182
Thermal Connection of the Hot Part to the Heat Sink	183
How to Avoid Heat Sinks	183
Pads as Heat Sinks	184
Use of Thermal Vias	184
Use of Fans and Filter Mats	185
Cooling Systems, Condensation, Peltier Elements	187
Creating a Thermal Prototype Early	187
Temperature Measurements	188
Thermal Simulations	191

6 Testing and Verification 192

Thinking About Testability from the Start	192
Basic Testability Measures	193
Preparing for Production Testing	194
About Those Test Points…	194
Tiny Holes to Save You Embarrassment	196
Ideas for Measuring Prototype Current	196
Collected Ideas for Improving Prototype Testability	197
Prototype Microcontroller Programming	198
Desktop Temperature Checks	199
Your Own EMC Tests Part 1	200
Your Own EMC Tests Part 2	204
PCB Markings	204
Pre-Production Series Lot Size	205
Tests Performed on the Pre-Production Series	205
Production Microcontroller Programming	210
Production Tests	211
Burn-in After Production	213

List of Abbreviations 214

Bibliography 216

Foreword

Today's available rapid prototyping ecosystems, which include a vast collection of code libraries and examples, seem to make hardware and firmware development easier than ever. This masks the fact that a working rapid prototype is far from production-ready.

Observing detailed electronic curricula around the world, I can conclude that rapid prototyping platforms are also ubiquitous in universities. This is not surprising. In fact, it is logical. Rapid prototyping platforms are excellent for demonstrating and studying the functional aspects of electronics without the time-consuming part of printed circuit board design. After all, universities have an obligation to teach their students the basics. How to create a production-ready version is a commercial aspect and off-topic.

There are many books on how to solve technical problems with electronic circuits. However, there are virtually no books on how to go from a working prototype to a production-ready printed circuit board that is robust enough to last in the field. This justifies a book that focuses on some of the best practices that lead to commercially successful electronics. As you read through the chapters, you will see that the secret to a production-ready version is not in its functionality. It is also about doing everything right: smart component selection, proper power supply design, electromagnetic compatibility, fast signal routing, thermal management, and testability.

This is the book I wished I had as a young engineer. But it has since become a book that appeals to experienced designers as well. Knowledgeable colleagues who read the book in draft form found much of what they already knew confirmed, but everyone also found a few things that were new to them. Therefore, I dedicate this book to everybody who develops electronics professionally, and I can assure you that it will be useful to everyone.

Matthew F. Berger
January 2024

Using this Book

This book contains many recommendations. To avoid disappointment, some comments on how to use them are necessary:

- The value of a particular recommendation often depends on the situation. A recommendation is best implemented as an option that can be omitted if there is no benefit.

- Often, it is more important to recognize potential flaws. Once you recognize a potential pitfall, the remedy is usually obvious. It may be the same as the recommendations in the book, or it may be something completely different.

- Do not use the diagrams with large question marks, such as in Figure 1 below. They are fundamentally flawed and are only displayed to properly illustrate a point.

- Until about 1969, there were hardly any rules of thumb in electronics design. Everything was analog and had to be precisely calculated or simulated. With the advent of TTL digital technology in that year, functioning electronic systems could suddenly be created without analysis at the lowest physical level. Knowledge of logic was enough. As systems grew, complexity increased and the physical level could no longer be ignored. But there was no time for in-depth analysis. People tried things out, which led to rules of thumb. Rules of thumb are not bad in themselves. However, it is very important to know the conditions under which they are valid. Therefore, rules of thumb mentioned here in the book are always accompanied by a comment about their scope of validity.

Although many of the statements are not derivable because they are based purely on experience, there are numerous references to literature. These are usually listed not as proofs, but as recommended literature for further study of the subject.

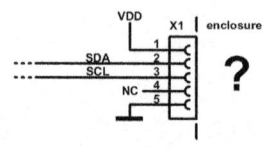

Figure 1. The big question mark indicates bad examples.

1 Component Selection

A successful electronic design is based on the right choice of components. The functionality of a component must match its requirements, but many other aspects should not be overlooked. For example, a connector is an electrically simple component, but its overall complexity is often underestimated. J.R. Barnes, an electrical engineer, kept statistics during his 13 years of design work at IBM; 90% of the problems in the field were related to connectors [R009]. How can we do better?

Connector Selection

The dumbest user applies excessive force and torque even when no force is required to make a connection. Surface-mount connectors are at acute risk of being torn off the printed circuit board, even if they are reinforced with adhesive. It is safe to say that the number one failure mode for a consumer device today is a misplaced surface-mounted USB jack. For this reason, always choose a through-hole technology or through-hole reflow connector with four pins. As a result, torque applied in any direction will not lift a connector off the printed circuit board.

For an IP-65 connector, i.e. one that is dust and water resistant, it is best to use a circular connector. This should have anti-twist protection and should not be connected directly to the board, as there is a risk of tearing and short-circuiting.

Avoid products that have two identical connectors with different functions. Instead, all functionally different connectors should have distinctly different looks. Even the dumbest user will try to make the wrong connection. Avoid this by choosing connectors that cannot be mated incorrectly, even with considerable force. For example, the old DSUB connector can still be a good second connector in addition to a USB connector.

How many times have you tried to plug a USB-A plug in the wrong way round? This is a particularly bad example of polarization. For the best user experience, do not use polarization, as with the USB-C connector. Of course, this reduces the number of usable pins on the connector by half. If you need all the pins to carry different signals, choose a connector whose polarization is clearly visible from a distance. In my opinion, an RJ-12 connector with a custom variant that shifts the nose to one side would make a good polarized connector.

4   Component Selection

It is advisable to assign supply voltages to sockets rather than pins, otherwise the probability of an accidental short circuit is high. This applies not only to the final product, but also to the often chaotic development environment.

Plugging and unplugging connectors during operation—known as "hot plugging"—requires special provisions. When disconnecting, the connector system must first disconnect the signals, then the power supply(s), and finally the ground. Ordinary connectors with the same pin types cannot meet this requirement because the disconnection order is random. If the power supply is disconnected first, but high-level digital signals are still connected, a connected receiving IC will be powered by the signals. The behavior of ICs that have no supply voltage but receive signal voltages at the pins is generally unpredictable and can lead to destruction.

A connector with a metal frame does not require proof of an electrostatic discharge (ESD) test with contact discharges to the pins. The discharge test can be performed with the ESD thumbtip alone. The resulting air discharge will always be directed to the metal frame due to its low impedance. This means that metal-frame connectors reduce the amount of ESD protection required on the PCB and can result in lower part count, as discussed in the Robust Interfaces chapter.

Do not specify holes for through-hole connectors that are too narrow. Otherwise, the solder will not penetrate well and you will have problems with both mechanical fixing and insufficient electrical contact, especially with high currents.

When electrical engineers talk about "derating," they usually mean choosing a margin for a given electrical value to make the circuit more robust. However, there are other types of derating. For example, assigning a function to each pin of a connector at the prototype stage is not wise. During development, multiple signal or power traces are usually added to the connector. If there is no spare, you have to change the connector. You may have to do this several times. Therefore, I recommend choosing a connector with about 10% more pins than you know will be used, or at least one spare pin.

Even with the finished product, it is not advisable to use fully occupied connectors. There is usually a follow-on product with additional features that often require additional connector pins. Replacing a connector because one or two more pins are needed is much more expensive in terms of layout redesign than choosing a connector with spare pins from the start.

In terms of electrical properties, there are two other important derating recommendations for connectors: no pin carries more than 50% of the rated current, and the connector operates no more than 25°C below the maximum temperature specified in the data sheet. The source for these recommendations is NASA [R001]. These are conservative values, but if you can, why not use them? If you have problems, you can point to the fact that you used tolerances like NASA's!

Pushbutton and Switch Selection

A hidden but important consideration when choosing a button or switch is bounce time. How long do they bounce in practice? A study of 18 different buttons and switches, each pressed 300 times in different ways, yielded the following results [R076]:

- The mean bounce time was 1.5 ms.

- The longest bounce time was 6.2 ms.

- Two switches fell out of this frame because they bounced at 11 ms and 160 ms, respectively. They were not included in the mean.

- Two switches of the same type had bounce times that differed by a factor of 2!

Therefore, it is a good idea to take a quick look at the bounce characteristics of a preselected pushbutton or switch type to make sure you have not accidentally chosen a type with a long bounce time. Buying at least three is advantageous to have a statistical basis for this purpose.

Clean software debouncing is more difficult than one might think. Consider a solution where the state of the switch is determined in a certain time interval that is much longer than the bounce time. For example, the state of the switch is determined every 200 ms if the maximum bounce time is 20 ms. This seemingly simple solution has important drawbacks:

- Disadvantage 1: The system reacts with a noticeable delay. All delays from 100 ms upward are certainly noticeable, and possibly those from 50 ms.

- Disadvantage 2: If the switch is operated quickly, the keystrokes are lost. Fast keyboard typists type a character well every 100 ms.

- Disadvantage 3: If there is interference at the microcontroller input at the moment the switch state is read, the wrong switch state may be determined, i.e. debouncing by waiting is not EMC safe.

Usually, these drawbacks are intolerable. More complex procedures are then required, but this is like jumping out of the frying pan into the fire. A more complex debouncing function must be revalidated and tested with every software change. This is because it cannot be ruled out that the latest changes to the code may affect the time-critical debouncing procedure.

Therefore, I recommend debouncing individual keys in hardware. The general rule of thumb is that debouncing in software is worthwhile for at least six to eight keys.

6 Component Selection

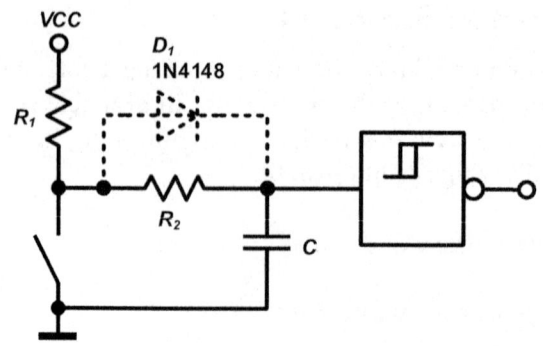

Figure 2. Debounce circuit for SPST switch.

Figure 2 shows the debounce circuit for an SPST pushbutton, i.e., for a single pole on/off switch. Notes on this circuit:

- R_2 is often omitted but be aware that without R_2 the capacitor is short-circuited. This stresses the capacitor and degrades the contacts of the switch due to high short-circuit currents.

- Although the switch-off time constant of $(R_1 + R_2) \times C$ is higher than the switch-on time constant $R_2 \times C$, the switch-on is delayed by the bounce, so that in practice the reaction times are approximately the same. Switching off does not bounce. If the switch-off time constant is to be shorter or the same as the switch-on time constant, a diode can be used, as shown in Figure 2. This short-circuits R_2 during charging, resulting in approximately VCC−0.7 V instead of VCC due to the voltage drop across the diode.

- For microcontroller inputs, check that they are Schmitt trigger inputs. This is important because the input level can sweep very slowly through the logically invalid range.

- FPGA inputs usually do not have a Schmitt trigger function, but this must be implemented explicitly.

- As a conservative assumption, assume a bounce time of 20 ms, or better yet, measure it.

$$R_2 = \frac{-T_{bounce}}{C \cdot \ln\left(\frac{V_{th,high-low}}{VCC}\right)} \qquad (R_1 + R_2) = \frac{-T_{bounce}}{C \cdot \ln\left(1 - \frac{V_{th,low-high}}{VCC}\right)}$$

Formula 1. Calculation of R_2 and C with $V_{th,high-low}$ the Schmitt trigger threshold from HIGH to LOW and T_{bounce} the estimated worst bounce duration. Calculation of R_1 and C with $V_{th,low-high}$ the Schmitt trigger threshold from LOW to HIGH.

Note that the digital input draws a leakage current, e.g., about 1 µA for the 74AHCT14 device. R_2 should be small enough to negate this leakage current, see Formula 1. Note too, that a capacitor is used as a timing element in this case. A C0G/NP0 ceramic capacitor should be used. However, if you want to do without one of these special capacitors select an X7R ceramic capacitor with a low tolerance. Electrolytic capacitors should be avoided due to their large tolerances. I always start with the capacitance when designing a debouncing circuit.

If you are using a conductive plastic switch, be aware that it can easily have a resistance of 200 Ω when closed. Such a switch may not bounce at all, in which case it can be connected directly to a Schmitt trigger input. Since the voltage changes slowly when the switch is closed, connect the switch only to a Schmitt trigger input.

Resistor Selection

Resistors are often deemed simple components—"cannon fodder"—but this should not hide the fact that they can also cause problems. A designer should always be aware that a resistor is not an R, a capacitor is not a C, and an inductor is not an L. Ask about the non-functional, parasitic properties of each element before using it. For example, it is important to know that with a chip resistor, there is lead inductance from the terminal caps into the interior and capacitance between the terminals. This results in the equivalent circuit shown in Figure 3, which represents a damped parallel resonant circuit. In principle, it is clear from this that a real resistor can also behave quite differently from an ideal resistor. However, because of the small size of a chip resistor, the parasitic properties—inductance and capacitance values—are expected to be very small. Are these not irrelevant in low-speed electronic circuits? Figure 4 shows that this is not the case. The parasitic capacitance of a common thick-film chip resistor with a high resistance value near 1 MΩ can have a considerable influence on the behavior of the component at frequencies below 10 MHz.

However, it is not advisable to select very high resistance values at all, as this will make the circuit too sensitive to electrical noise and fail the EMC test. It is better to keep values below 1 MΩ. Therefore, the general rule of thumb is that chip resistors up to 10 MHz behave almost ideally.

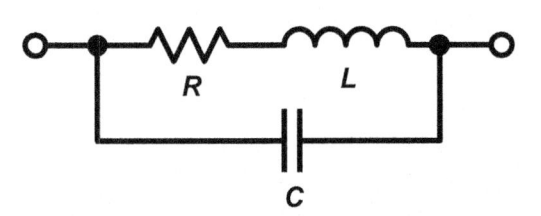

Figure 3. Equivalent circuit diagram for a chip resistor.

8   Component Selection

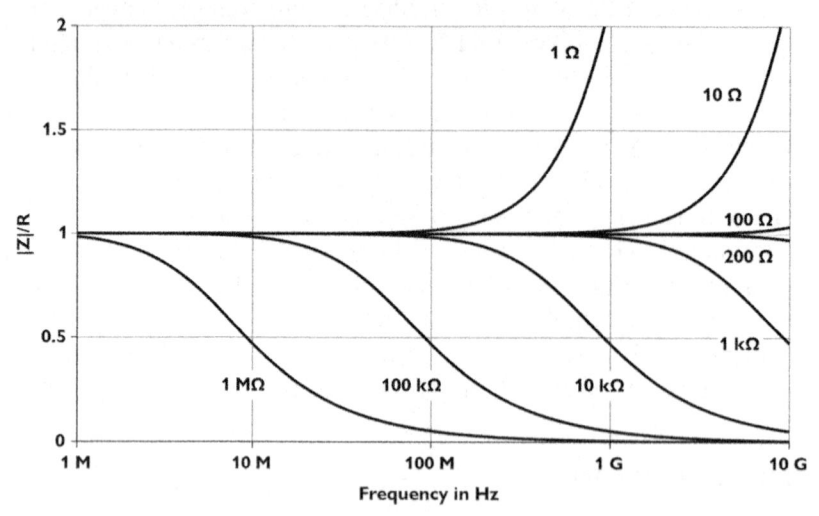

Figure 4. Magnitude of impedance normalized to DC resistance value vs. frequency for size 0603 thick-film resistors, based on R086 and R087.

Changing to thin film resistors hardly changes the curves. At most, it worsens them due to higher parasitic inductance when the thin film is laser trimmed. Switching from a 0603 to a 0402 package reduces the parasitic capacitance by a third, but only increases the natural resonant frequency by a factor of 1.2, so there is little gain. Switching to a flip chip—a resistor that has no caps, only pads—also improves the situation, but not useful enough [R087].

If your circuit is sensitive to resistor tolerances, consider using resistor arrays. The ratio of the resistors in the array has a lower specified tolerance than the individual values. A resistor divider using a resistor array will better maintain the ratio from device to device and over the full temperature range. Different resistor values in the same array are also available. Resistor arrays are space-saving when used to provide digital buses and other parallel multiple traces with pull-up and pull-down resistors.

However, there is no cost savings with resistor arrays because individual resistors are inexpensive to populate. In addition, there are significant drawbacks: array resistors have a lower maximum power, a lower maximum voltage, and a more constrained layout. They do not offer space savings due to less flexibility if used for individual signals rather than for buses.

Finally, note that a 0 Ω resistor cannot carry an infinite current because it has a residual resistance greater than 0 Ω. A 0 Ω resistor of size 0603 typically has a maximum current capability of 1 to 2 A. Refer to the data sheet for the selected type and be aware of this limitation.

Potentiometer Selection

Potentiometers have some advantages, such as inherent linearity (for linear potentiometers), non-distortion, and non-volatile settings. Otherwise, they can cause real problems in professional designs:

- Potentiometers are large and full of parasitic effects. Therefore, they have very limited use in AC signaling, unless the product of signal frequency and resistance is less than $1 \text{ MHz} \times \Omega$.

- They are time consuming and expensive to assemble and test.

- Potentiometers still have a "final resistance" even at the stops; it is not possible to set them exactly to zero ohms.

Nowadays it is common practice to use encoders instead of potentiometers and to evaluate their signals with a microcontroller.

Trimmer Selection

The typical application of trimmers is in the zero offset adjustment of op-amps. Today, the offset is in the sub-millivolt range and can be tolerated. Fortunately, there is no longer a standard application for trimmers. In fact, for a quality product, it is almost unthinkable today that the accuracy of the device depends on a manual adjustment step. With today's digital systems, you can replace a trimmer with a digital potentiometer. This may even allow the system to calibrate itself, which would be a significant quality improvement. However, if a trimmer is to be used, the negative points mentioned above for the potentiometer should be considered, as they also apply to a trimmer. In addition, the following points specific to trimmers must be observed:

- Production costs increase with each manual trimming step.

- Trim resolution is only theoretically infinite; precise adjustment to a point is impractical. It is better to use a fixed resistor plus trimmer for fine adjustment, or a spindle trimmer for fine adjustment over several turns.

- Carbon trimmers with a tolerance of 20% have no use today.

- Cermet, or "CERamic Substrate with METal film deposited" trimmers are one possible component that can be used. They range from $10 \, \Omega$ to $2 \, M\Omega$, with a tolerance of 10%. Smaller designs with lower intrinsic capacitance are available for higher frequencies. In general, a trimmer is not a high frequency component!

- Pay attention to the accessibility of the trimmers. Use flat trimmers in the center of the board and upright trimmers at the edges.

Capacitor Selection

Like a real resistor, a real capacitor does not remain capacitive to arbitrarily high frequencies. Figure 5 shows the typical impedance characteristics of capacitors as a function of frequency. Initially, at very low frequencies, the capacitors behave as they should: at ten times the frequency, the impedance drops by a factor of ten. However, this behavior ends at a certain frequency. There, a more or less pronounced self-resonance occurs. Above the resonant frequency, the capacitor begins to behave inductively, and the lead inductance dominates and determines the impedance. The component can no longer be used as a capacitor, i.e. the phase response is exactly 180° inverted. Within a filter, this behavior is disastrous; the filter no longer works as it should. If it is an active filter, the phase reversal can cause positive feedback and thus spurious oscillations.

However, for power supply decoupling, the phase is not important—only the impedance value. For example, the 10 nF capacitor has the same low impedance of 1 Ω at 200 MHz as it does at 20 MHz. As a power supply decoupling capacitor, its effects are the same at both frequencies. See the power supply design section for more information.

The minimum of the impedance curve corresponds to the equivalent series resistance (ESR) of the capacitor. The ESR results from losses during charging or recharging of a capacitor caused by reorientation of the molecules. Table 1 shows the typical ranges of ESR values for different types of capacitors. When a capacitor becomes part of a control loop, its ESR value must meet the requirements. Low-drop-out

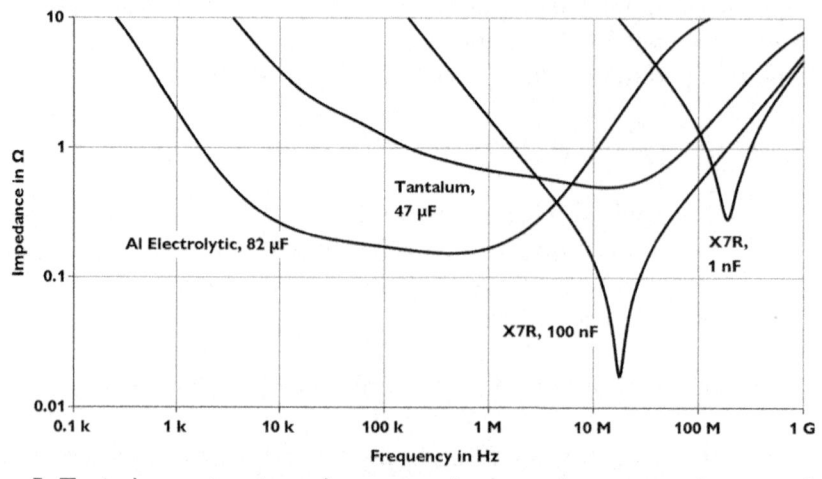

Figure 5. Typical capacitor impedance magnitude vs. frequency: aluminum liquid electrolytic capacitor 380LQ820M450H022 with 82 µF in 22 mm diameter and 30 mm long case. Tantalum electrolytic capacitor 16TAJA475K0RNJ with 4.7 µF in A-type case, 100 nF ceramic capacitor 0603YC104KAT2A in 0603 case (X7R), 1 nF ceramic capacitor 0201YC102KAT2A in 0201 case (X7R).

Type	ESR
Liquid Electrolytic Capacitors	0.1–10 Ω Increases with frequency, age, temperature, and ripple amplitude.
Tantalum Electrolytic Capacitors	0.1–10 Ω, Increases with frequency.
Ceramic Capacitors	0.01–0.5 Ω, relatively stable over all of the above parameters.
Film Capacitors	0.01–0.1 Ω, relatively stable over all of the above parameters.

Table 1. Typical ESR values for capacitors. Because of the wide range, the actual ESR value of the capacitor to be used must be found in the data sheet.

regulators (LDOs), for example, demand that the ESR value of the output capacitor be within a certain range. In most cases, you cannot simply replace a tantalum cap with a cheaper ceramic cap. Then, you must also change the LDO type.

Liquid electrolytic capacitors have a good capacitance/cost ratio, otherwise they suffer from a number of disadvantages that are worth mentioning:

- They have comparatively low lifetimes. They are electrochemical devices that dry out over time.

- They have a low resonant frequency.

- They have a relatively high ESR, which is often the limiting parameter for their use, especially within supplies. Otherwise, the capacitor becomes too hot, which reduces its short lifetime even more, or it simply overheats.

- All electrolytic capacitors are polar elements, and there is a risk of incorrect placement. This is aggravated by the fact that such an error cannot be detected by any electrical test or burn-in time. The only way to detect misplacement is by visual inspection. Wrongly polarized electrolytic capacitors suffer from a reduced lifetime. Devices will fail before reaching their guaranteed lifetime. Correctly placed electrolytic capacitors must not be subjected to high negative voltages during operation.

- Electrolytic capacitors are large components, which means they must be mechanically and securely mounted to the printed circuit board using either a through-hole device or an SMD device glued to the printed circuit board.

- Liquid electrolytic capacitors have a high leakage current.

- Liquid electrolytic capacitors have a limited shelf life. When purchasing, make sure you are not buying an aged batch.

- The predetermined breaking point must be oriented such that it does not point in the direction of an inductively activated element (e.g., a relay or a fan). The capacitor must be sufficiently far away so that, in the event of outgassing, there is no ignition by a spark.

- Liquid electrolytic capacitors boil in a vacuum. They cannot be used in satellites.

- Today, liquid electrolytic capacitors are really only a good choice where high capacitance (i.e., 100 µF and above) is required.

Tantalum capacitors have a solid electrolyte, which solves some of the problems of liquid electrolytes. However, tantalum capacitors have their own significant drawbacks, including the tendency to fail under reverse voltage, surge voltages, high ripple currents, and high inrush currents. Tantalum is also a rare element. In the past, there have been bottlenecks in the availability of tantalum. Today's wisdom is to avoid them whenever possible. Fortunately, the typical applications for tantalum capacitors are increasingly being replaced by ceramic capacitors. Five 10 µF X7R ceramic capacitors in the 0603 package with 6.3 V dielectric strength can now be purchased for as little as 20 cents each. In contrast, five tantalum capacitors with the same capacitance and dielectric strength cost at least 30 cents each. Often a tantalum capacitor is used out of habit. For any tantalum capacitor, ask yourself if it really needs to be a tantalum type.

Aluminum polymer capacitors have been available for some time. Compared to aluminum liquid electrolytic capacitors, they have lower ESR and ESL values and longer life. They are not easily ignited by overvoltage, incorrect polarization, excessive ripple current or inrush current. However, aluminum polymer capacitors tend to be more sensitive to vibration and are larger for the same capacitance and dielectric strength.

Ceramic capacitors have none of the disadvantages mentioned for electrolytic capacitors, but they do have their own, sometimes serious, disadvantages, depending on the type. A three-character code divides ceramic capacitors into categories. The first character indicates the lowest temperature at which the capacitor can be operated. For example, the letter X, as in X7R, corresponds to -55°C. The second character defines the maximum temperature, in the case of 7 it is 125°C. The third character describes the maximum amount of capacitance change over the operating temperature range, e.g., R stands for ±15%. These categories are divided into Class 1 and Class 2, see Table 2. Class 2 ceramic capacitors are piezoelectric. Vibrations can produce millivolt voltages at their terminals. If they are used in the signal path, the device may become susceptible to microphonics, i.e., the vibrations appear as patterns in the signal. Use Class 1 capacitors unless the signal level is sufficiently high and the resolution is low.

The effective capacitance of Class 2 capacitors is highly dependent on the operating conditions. In particular, as shown in Table 3, the operating voltage has a significant effect on the effective capacitance. Ceramic capacitors are typically measured at 1 V RMS. At higher operating voltages, Class 2 capacitors suffer a massive reduction in capacitance. On the other hand, tantalum capacitors have only about 2% less capacitance when operated at nominal voltage [R192].

Wherever you use ceramic capacitors, be aware that their low ESR and therefore high Q factor, together with their inductance, make them susceptible to oscillation. An example is the resonance created by the trace inductance and the decoupling capacitors excited by hot plugging. In general, ceramic capacitors with particularly low ESR values should not be selected.

All ceramic capacitors are sensitive to mechanical stress. If the PCB bends during assembly or operation, ceramic capacitors tend to break, always causing a short circuit. In the automotive industry, for example, it is common practice to connect two capacitors in series instead of

Subtype	Description
Class 1 C0G, NP0	Capacitance values stable against voltage, frequency and temperature. Low losses. Application: Filters, timers and high frequency circuits.
Class 2 X5R, X7R, Y5V	Up to 1 µF in a reasonable package size. Higher capacitance at the expense of variability with voltage, frequency, temperature and age. Piezoelectric vibration and shock can induce voltages! A few mV occur with slight vibration. Application: Decoupling capacitors, coarse filters and coupling capacitors for relatively high-level signals.

Table 2. Typology of ceramic capacitors.

14   Component Selection

Type	C nF	V_{max} V	Size	F_{res} MHz	ESR mΩ	ESL nH	% @ 3.3 V/ % @ V_{max}	Max ΔC to 125°C in %
X7R	1	16	0603	193	268	0.68	-3.2/-41	-5.8
X7R	1	16	0201	285	256	0.32	-0.7/-6.1	-10.5
X7R	100	16	0603	17.6	19.0	0.77	-1.5/-16	-12.1
X7R	100	50	0603	17.7	19.4	0.81	-1.5/-66	-11.4
X7R	100	16	0402	18.7	15.7	0.73	-5.8/-55	-14.7
Y5V	100	50	0603	16.5	22.0	0.93	-16/-85	-70

Table 3. Representative values for multilayer ceramic capacitor characteristics, X7R with ±10% factory tolerance and nickel/tin terminals, Y5V with +80%/-20% factory tolerance. Based on data supplied by AVX.

one capacitor, with one of them placed at a 90° angle, see Figure 6. If one of the capacitors breaks because the board is bent towards it, the other one will still work. But even with such a measure, you should never place ceramic capacitors at the edge of the board, but as far inside as possible.

As an alternative, or as an afterthought, there are so-called soft-termination capacitors, which are much more tolerant of bending. However, because of their higher cost than standard capacitors, they may not be feasible in a high-volume product. There is a latent risk that someone will replace them with regular capacitors due to lack of knowledge in order to save money.

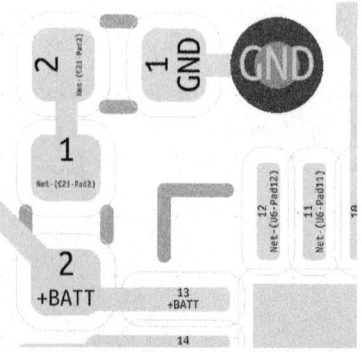

Figure 6. Part of a layout with two ceramic capacitors in series at a 90° angle.

Selecting Power Inductors, RF Coils and Chokes

Real inductors have limitations and tradeoffs depending on the type. The three main categories of inductors are:

- Power inductors optimized for high currents, but large SMD components compared with others.

- RF coils optimized for high bandwidth and small size but having a comparatively large DC resistance.

- Chokes for low-loss at DC but high-loss at RF frequencies.

Figure 7 shows the impedance diagram for selected inductors representing these three categories. The application usually dictates which category to select, e.g., power inductors for DC/DC converters, RF coils for high frequency filters, chokes for harmonic suppression. However, it should be noted that the best choice is not to use an inductor at all. There are many reasons for avoiding inductors:

- Inductors are less standardized than capacitors, coming in more different footprints. This reduces the number of alternatives once a specific inductor is selected.

- Inductors tend to be comparatively expensive because of their more complex 3D structure compared to capacitors.

- Inductors have a DC resistance that can be well into the 1 Ω range. This makes it difficult to build low-pass or band-stop filters.

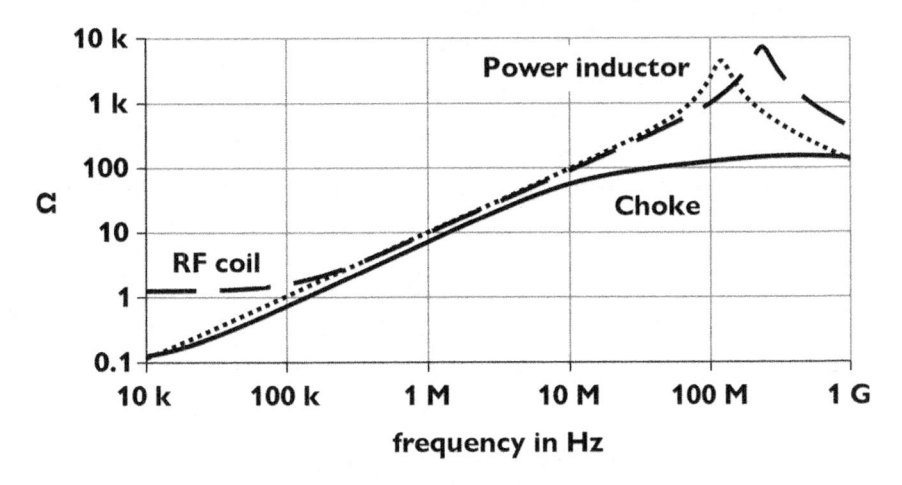

Figure 7. Frequency response of the 1.5 µF RF coil LQW15CE1R5M10, of the 1.5 µF power inductor LQH32PN1R5MNC and of the choke BLM18RK121SN1, all from Murata.

16 Component Selection

- Inductors usually contain a ferrite core, which is brittle. There is no defined failure mode as there is with ceramic capacitors, which always results in a short circuit when broken. Broken inductors can result in an open or short circuit. Unlike ceramic capacitors, there is no point in placing two inductors in series at a 90° angle (see the section on capacitors just above). Therefore, avoid using inductors in equipment that is subject to vibration or shock. As a workaround, consider using soft-termination inductors if they are not too expensive.

- Inductors with coils can saturate if the current is too high. However, controlling current is more difficult than controlling voltage in voltage-based systems.

- Above the Curie temperature, the ability to magnetize a coil core is lost and it effectively becomes an air coil. Manganese-zinc ferrites have Curie temperatures as low as 120°C.

- When processing frequencies in the range of 20 Hz to about 1 kHz with a coil, coil whine, an audible, potentially disturbing buzzing of the element, can occur. The magnitude of this magnetic noise depends on structural resonances. Replacing the humming coil with a different type with the same inductance but a different size may solve the problem.

- When you hit a magnet, it loses its magnetization. However, I have never heard of vibrations causing induced voltages in coils with ferrite cores. However, the ferrite material is brittle and there is a latent risk of a ferrite core breaking apart. This could be counteracted by generously applying adhesive to the coil or by casting the circuit, but neither method is usually an option.

- Coils in the nanohenry range are unfounded, since a 0.2 mm wide stripline, together with the return in a ground plane 0.2 mm below, already has an inductance of about 6 nH/cm. A more detailed analysis of trace inductance can be found in the chapter on power supply design.

Example methods to avoid the use of coils are:

- If you use a linear regulator instead of a converter, you can save the power inductor and choke(s) of the input/output filter(s).

- Use an active filter instead of a 2nd order passive one or a stripline filter for very high frequencies.

Additional Passive Component Considerations

Passive components come in specific sizes; Table 4 provides an overview. The cheapest values are usually those of the smallest E-series: 10, 15, 22, 33, 47, and 68.

Table 5 lists some conservative derating approaches for passive components. Why not follow them, if possible, then you can say that you derated your components according to NASA!

Placement machines have loading capacities ranging from 15 to 120 reels, depending on the machine. The smaller the number of surface-mount components in a circuit, the more flexible the assembler is and the faster they can proceed with the placement. If they can work with a smaller machine, it is also possible for the assembler to compute more tightly. At the end of each schematic revision, go through the BOM and analyze the potential for simplification. Often, for example, there is no reason to use two different value pull-up resistors, such as 10 kΩ and 47 kΩ, and it is sufficient to work with one value.

Size	Comment
0402 (1 mm to 0.5 mm)	Printing not available on resistors. Higher or lower assembly costs (depending on the assembler). Often 1/16 W $\rightarrow$ Rmin = 600 Ω (50% margin) at 5 V. (!)
0603 (1.5 mm to 0.75 mm)	Can be easily assembled by hand. Printing available on resistors. Often same power dissipation for resistors as 0805.
0805 (2 mm to 1.25 mm)	Considered too big today. Not much cost advantage over 0603.
1206 (3 mm to 1.5 mm)	For higher power dissipation with resistors.

Table 4. Advantages and disadvantages of SMD resistor sizes. Size code example: 0402 = 40 mil × 20 mil, where 1 mil = 0.001 inch = 0.0254 mm.

Part	Operation at:	Source
Connector	Up to 50% of maximum current.	NASA [R001].
	Up to 25°C below maximum temperature.	NASA.
Switch	Up to 75% of maximum current for resistive loads.	Arsenault [R197].
	Up to 75% of maximum current for capacitive loads.	Arsenault.
	Up to 40% of maximum current for inductive loads.	Arsenault.
	Up to 20% of maximum current for motors.	Arsenault.
Resistors	Up to 60% of maximum power.	NASA.
	Up to 80% of maximum voltage.	MIL-HDBK-1547.
	Up to 60% of pulse power.	MIL-HDBK-1547.
Capacitors	Electrolyte: up to 80% of maximum voltage.	MIL-STD-198E.
	Tantalum: up to 50% of maximum voltage.	MIL-STD-198E.
	Tantalum: up to 70% of surge current.	HIL-HDBK-1547.
	Ceramic: up to 50% of maximum voltage.	MIL-HDBK-1547.
	Ceramics: up to 70% of surge current.	MIL-HDBK-1547.
Coils	Up to 50% of maximum voltage.	NASA.
	Up to 60% of maximum temperature.	NASA.
	Up to 30°C below insulation maximum temperature.	MIL-HDBK-1547.
	No data in cited works about saturation current.	

Table 5. Examples of derating for passive elements.

Part	Operation at:	Source
Diodes	Up to 70% of blocking voltage.	MIL-HDBK-1547
	Up to 50% of forward current.	MIL-HDBK-1547
	Up to 50% of surge current.	MIL-HDBK-1547
	Up to 50% of power loss.	NASA [R001]
	Junction temperature not above 110°C.	NASA
LED	Up to 80% of forward current.	Arsenault [R197]
MOSFETs	Up to 75% of drain-source voltage.	MIL-HDBK-1547
	Up to 75% of gate-source voltage.	MIL-HDBK-1547
	Up to 75% of drain continuous flow.	MIL-HDBK-1547
	Up to 50% of nominal power.	NASA
	Junction temperature not above 110°C.	NASA
OpAmps, Regulators	Up to 80% of supply voltage.	NASA
	Up to 75% of power loss.	NASA
	Barrier temperature not above 100°C.	NASA
Digital ICs	Up to 90% of drive current.	Arsenault
	Up to 75% of maximum frequency.	Arsenault
	Up to 75% of power loss.	NASA
	Junction temperature not above 100°C.	NASA
	At least 110% assumed propagation delay.	MIL-HDBK-1547

Table 6. Examples of derating for active elements.

Semiconductor Selection

We will now discuss the selection of active components (i.e. semiconductors). For these, the margins you deliberately add are especially important because of the sometimes large tolerances of the parameters. Table 6 lists some derating approaches.

When selecting components in general, and semiconductors in particular, never use "typical" values for design calculations. Always use guaranteed minimums or maximums. The appendix to R084 aptly states that "typical" can mean the following:

- "At least one manufactured IC complies with this specification."

- "Half of the ICs in production are better, the other half worse."

- "There was once a time when half of the ICs in production were better, and the other half worse."

- "Most ICs produced are close to this specification, some are better and some are much worse."

Problems with transistors and diodes are often specific to a particular type. A treatise on this is beyond the scope of this book. However, some general hints are possible, and these are mentioned here.

Diodes and LEDs

I know of an automotive electronics manufacturer that prohibits the use of diodes in glass packages. Incident light can turn the diode into a current source, similar to a solar cell.

If you want a fast-reacting freewheeling diode, you must not use an ultra-fast recovery diode, which has a very fast turn-off time and therefore a very long turn-on time. What you need is a diode with a very fast turn-on time, but a slow turn-off time. Therefore, 1N4148 or 1N4007 is a good choice for a fast turn-on diode.

If you are selling more than one unit per customer, consider whether you want to supply all the LEDs with current sources and specify the so-called LED bin. The luminosity of the same type of LED varies so much even within a single production batch that they light up visibly differently when driven by the same voltage and current limiting resistor. This can give customers the impression that the quality of the devices is poor. Manufacturers therefore regularly offer LEDs on reels, sorted by luminous intensity in relation to current. However, distributors usually do not offer the option of ordering LEDs from a specific bin to make things easier for them, but this makes things more complicated for you. Contact the distributor if you want to get LEDs of the same type and all from the same bin.

If you use light pipes, make sure they are ROHS compliant if they have pins and are soldered to the PCB. Anything soldered to a PCB must be ROHS compliant, even if it has no electrical function.

Transistors

If you use half-bridge transistors with freewheeling diodes between the source and drain, these diodes will burn out instead of the transistors' internal substrate diodes in the event of a mismatch. Diodes are usually easier to solder out and back in than transistors.

Bipolar transistors cannot be connected in parallel, but metal-oxide-semiconductor field-effect transistors (MOSFETs) can, see the section on thermal management.

If you realize a current source with a MOSFET, it will only work from a certain minimum voltage, make sure that its offset is acceptable.

Optocouplers

Optocouplers in a white housing are those without a black casing. The black casing is usually desirable and standard for ICs, as incident light cannot interfere with the circuit and heat is better radiated. However, the black coating leads to a higher capacitance between the input and output circuit of the optocoupler due to its higher permittivity compared to air. This is why optocouplers in a white housing are 1.5 to 3 times faster than the same type in a black housing. However, do not use white optocouplers in areas that can be exposed to light.

Operational Amplifiers

Every characteristic of an operational amplifier (op-amp) is a compromise between all the parameters of this type of device. A random or historical choice is unlikely to provide the best solution. Therefore, it is worthwhile to have a precise initial op-amp specification. If your op-amp search returns a large number of choices, you probably have not specified your requirements in enough detail. Most of the time, if you know exactly what you want, the situation is reversed: you do not find an exact match.

I start with the most important criterion, then add the next most important, and so on. In this way, the selection of suitable types is gradually narrowed down until only a few components are left to choose from. The selection is often easier on the manufacturer's web site than on the distributor's web site, since the former sometimes offers categorizations such as "high speed" and "low power," and the required parameter range can be better defined.

The "rail-to-rail" property is an op-amp parameter that will be discussed here. Conventional bipolar types derived from the original "mother of all amplifiers" µA741, such as the LM318, were designed for a ±15V supply. Typically, they achieve an output swing of ±13 V, i.e., they need a reserve of about 2 V for the supply voltages. However, there is still a considerable useful output voltage range. Some of these operational amplifiers can be operated from a single 5 V supply, but there is virtually no output voltage range left. At 5 V or lower, an "output rail-to-rail" op-amp type is required. Op-amps with rail-to-rail output characteristics require virtually no supply voltage and ground reserves. However, it should be noted that the output rail-to-rail characteristic depends on the load at the op-amp output. Most rail-to-rail op-amps only reach the supply voltage or ground with an open output. Any load reduces the output voltage range by the ratio of the load impedance to the output impedance as a voltage divider.

An input rail-to-rail characteristic may also be required if the entire input voltage range from ground to supply voltage must be covered, as may be required for a voltage follower.

Multiple op-amps in the same IC package often have a total quiescent current that is only slightly higher than that of a single op-amp. On the other hand, there are some disadvantages:

- Inflexibility in power supply and PCB layout.

- Thermal and supply voltage interactions.

- High-frequency crosstalk to adjacent traces, making a high internal isolation between the op-amps useless.

- Different pinouts for different multiple-op-amp ICs.

- Do not forget to set unused op-amps in a defined state.

My choice is to use multi-op-amp ICs only in low-power devices.

Advantages	Lower offset current. Lower cost. Many op-amps to choose from.
Disadvantages	Comparatively slow. No hysteresis (without wiring). Possible instability due to stray capacitance feedback.

Table 7. Comparison between a rail-to-rail output operational amplifier that acts as a comparator and a dedicated comparator.

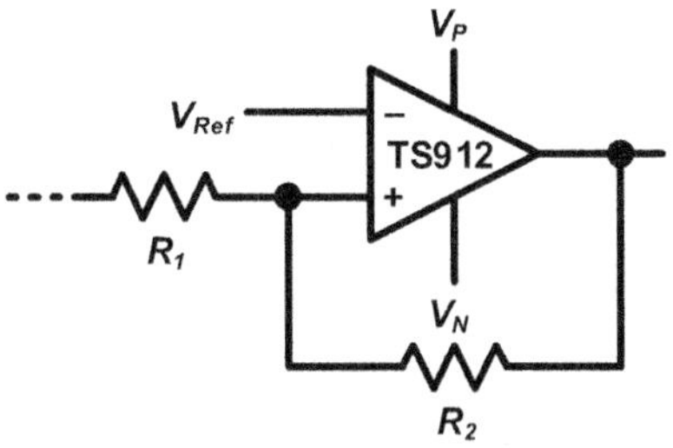

$$\Delta V_{hysteresis} = \frac{\Delta V_{out}(R_1 + R_2)}{R_1}$$

$$V_{LH} = \frac{R_2 \times V_{Ref} + R_1 \times V_P}{R_1 + R_2} \qquad U_{HL} = \frac{R_2 \times V_{Ref} + R_1 \times V_N}{R_1 + R_2}$$

Figure 8. Operational amplifier as comparator with hysteresis. Hysteresis $\Delta V_{Hysteresis}$ due to output voltage range and resistors R_1 and R_2. Note that the signal applied to R_1 must come from a source with much lower impedance than R_1 (or use the source impedance if it is sufficiently stable). Upper switching threshold V_{LH} and lower switching threshold V_{HL} due to positive supply voltage V_P, negative supply voltage V_N, reference voltage V_{Ref}, and resistors R_1 and R_2.

Output rail-to-rail op-amps are very similar to comparators; however, there are differences, which can be significant depending on the application, as shown in Table 7. Stability problems due to capacitive feedback can be solved by applying a bit of positive feedback to the op-amp, as shown in Figure 8.

Reference Voltage Sources

Single-supply operational amplifier circuits often require a reference voltage midway between VCC and ground, which can easily be provided by a voltage divider as shown in Figure 9. However, the smallest disturbances at VCC are directly visible in the signal. In addition, there is a risk of feedback due to the current drawn by the operational amplifier and thus modulation of the local VCC potential, or due to the current drawn by the next stage.

A better solution is to use an inverting amplifier as shown in Figure 10. An additional capacitor C_1 can be used to remove noise from VCC and keep the reference voltage clean. The value of C_1 is determined by the lowest frequencies to be filtered. The time constant of the filter is

$$\tau = C_1 \times (R_1 \times R_2) / (R_1 + R_2)$$

There is still a positive feedback path via VCC. Varying current draws of the op-amp itself or the subsequent stage(s) can change the local VCC potential; this change is followed by the reference voltage if it is low enough in frequency. It may be possible to solve the problem by lowering the filter's cutoff frequency, but the time constant may become unacceptably large and the system may take too long to set up.

24 Component Selection

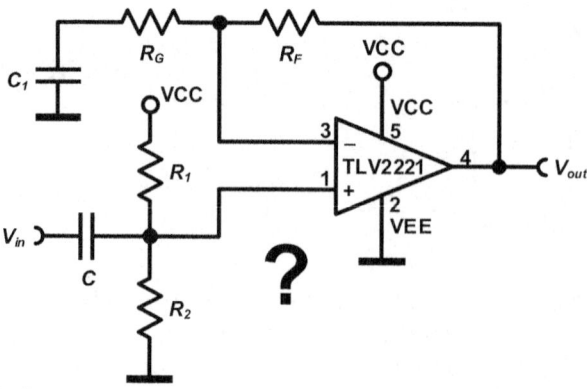

Figure 9. Single-supply AC op-amp circuit in which the potential at the positive op-amp input is defined by VCC, ground, and a voltage divider.

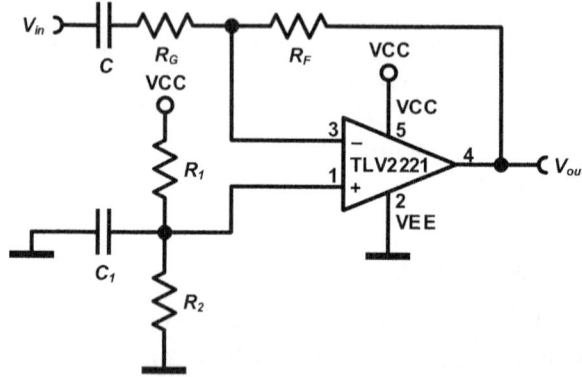

Figure 10. Inverting single supply op-amp circuit for AC voltage signals. The reference potential is defined by a voltage divider and stabilized by C_1 [R071].

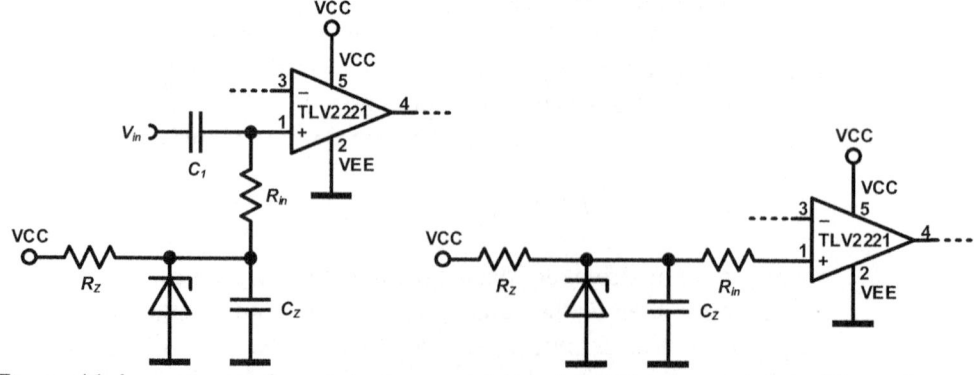

Figure 11. Inverting single supply op-amp circuit for AC voltage signals. The reference potential is defined by a Zener diode and stabilized by C_Z [R071].

Further notes on Figure 10:

- The accuracy of a reference voltage obtained with a two-resistor voltage divider of 1% is up to 2% inaccurate, depending on the ratio of the values.

- Temperature compensation is achieved by using a resistor array.

- In general, voltage divider references are too noisy for low-noise design, so switch to the other types of references.

- If it must be a non-inverting amplifier, see R071. If the input signal must be raised to a level, not just amplified, see R073.

A disadvantage of the solution shown in Figure 10 is that it is not frequency independent. As an alternative, what about using a Zener diode to generate the reference voltage, as shown in Figure 11? This is a typical theoretical solution because of the following practical problems:

- The breakdown voltage of low voltage zener diodes is highly dependent on current and temperature.

- Zener diodes are noisy, you need to add a capacitor in parallel, this may be physically too large for a low noise circuit.

Professional designs no longer use zener diodes to provide reference voltages. Use a bandgap reference IC or a linear regulator.

Some notes on bandgap reference voltage ICs:

- They are highly accurate, space and power efficient. For example, the LM4050, 2.5V, available in the SOT-23 package, operates from 65µA, has a temperature coefficient of 50 ppm/°C and an initial accuracy of ±0.1%. This is a voltage reference!

- The cost of a reference depends on the temperature coefficient and tolerance, from the cheapest to the most expensive can be a factor of 1:3.

- Fixed references only possible with certain bandgap voltages: 1.2 V, 2.5 V, 5 V, 10 V, etc.

- Variable references require a voltage divider circuit, the accuracy of the voltage reference is reduced by the resistor tolerance.

- The load impedance must be significantly greater than the resistors of the reference circuit, i.e. at least a factor of 10. Otherwise, the reference must be provided with a voltage follower. If so, pay attention to the temperature response of the op-amp used for the voltage follower: it may be worse than that of the reference! The same goes for noise.

- If the supply voltage drops, the reference voltage is stays at its value and does not follow. This can lead to one-sided clipping.

- There are many different references, but little standardization, some need a decoupling capacitor, some do not, the pin-out is different, etc., see datasheet. The bandgap device is sometimes shown as a zener diode in schematics, but it is an IC.

Consider using a linear regulator instead of a bandgap IC. The former has the following advantages:

- Very good reference voltage even with a very noisy supply voltage

- Low inherent noise and low output impedance, therefore very suitable for low-noise designs.

- Only clean solution for a 1.65 V reference voltage for, e.g., a 3.3 V op-amp circuit.

However, a linear regulator IC typically requires more board space than a bandgap regulator IC.

Choosing Integrated Circuits for Simple Logic Tasks

Sometimes a simple logic operation, even if it is only an AND gate, needs to be implemented in hardware. In my experience, single-gate ICs can have unusually high supply currents, contrary to the data sheet. The reason for this is unknown. Check this if you have a low power device. If you are using a multi-gate logic IC, but only need one gate, make sure the others are in a defined state, otherwise they may oscillate, increasing the current draw even more.

While we are still talking about simple logic and implementing more than one gate, let me mention complex programmable logic devices (CPLDs). CPLDs have large non-volatile memories and instant startup.

Memory Selection

Table 8 provides an overview of common external storage media for microcontrollers. Flash memory is ubiquitous today and is measured in terms of memory volume and affordability. However, they have a comparatively low number of write cycles. This limit is quickly reached with frequent writing (e.g. continuous recording of sensor data). If you write every second, the write cycle limit of Flash memory in Table 8

Parameter	Flash Memory	EEPROM	FRAM
Example type	LE25U20AMB	M24M02-DR	FM24W256
Write cycles	10^5	4×10^6	10^{14}
Power down	10 µA	5 µA	15 µA
Maximum clock	30 MHz	1 MHz	1 MHz
Write cycle*	max. 5 ms	max. 5 ms	max. 300 µs
Memory	256 k x 8 bit	256 k x 8 bit	32 k x 8 bit
Data retention	20 years	200 years	150 years
Interface	SPI	I2C	I2C
Indicative price	See daily price	≈ 2 x flash price	≈ 10 x flash price

* Flash: block write (256 bytes), EEPROM: page write (256 bytes), FRAM: 256 consecutive bytes

Table 8. External non-volatile memory for microcontrollers.

will be reached in 3¼ years. This is not a long life for an industrial system. However, cheap consumer Flash cards may only have 500 write cycles! Make sure that the specified number of write cycles is compatible with your product requirements.

Another surprise with Flash memory is the completely different access times. Using wear leveling and variable assignment of the logical address to the physical memory location, the manufacturers try to prevent an uneven write frequency of the individual memory cells to bring the number of write cycles into the range shown in Table 8. This algorithm is not transparent to the user and takes more time if the Flash has already been written to. A new Flash memory may excite you with fast write cycle times. However, be sure to calculate the maximum specified write cycle time at the outset. Otherwise, a device will fail erratically in the field. The failure source will be difficult to determine.

With a classic EEPROM—Flash memory is, strictly speaking, also an EEPROM—you do not have the above problems. Classic EEPROMs have no hidden algorithm. You may think they are slow, and this is true when writing single bytes. For the M24M02-DR device, writing 256 bytes individually would take about 1.3 seconds. However, when using the page write function, the write times for EEPROMs and Flash memories are comparable, as shown in Table 8.

28 Component Selection

FRAM should be mentioned as an alternative to Flash memory and classic EEPROMs. It has a very high number of write cycles combined with a very short write cycle time. Unfortunately, the price of this type of memory is quite high.

Analog-to-Digital Converter Selection

Today's microcontrollers usually have an integrated analog-to-digital converter (ADC). Its resolution often does not seem exciting. However, if you want a higher resolution, be aware that the resolution limit can be set by the noise in the system, so a high-resolution conversion will result in a lot of meaningless data (see Table 9). For 12-bit resolutions and higher, it is challenging to design the analog part to sufficiently reduce the noise level. See R078 for practical layout recommendations.

Digital-to-Analog Converter Selection

For digital-to-analog converters (DACs), the question is: What tools should I use?

- DAC of the microcontroller itself.

- DAC with pulse-width modulated (PWM) output or pulse code modulation (PCM).

- DAC with specific ICs.

- DAC with discrete circuits.

This question arises because microcontroller DACs are often poorly specified in terms of DC voltage if they are primarily intended for audio output. An absolute, accurate level plays a secondary role in audio applications. The question also arises when the microcontroller's DAC port is occupied by something more important, when high speed is required, or when a microcontroller has no DAC at all.

If the microcontroller does not have a built-in DAC, or the pin is occupied by something more important, a PWM solution may be possible. You only need one pin, and most microcontroller hardware is prepared for PWM output. However, a DAC using PWM has some drawbacks:

- They usually require an analog filter.

- A high clock frequency is required for a clean sine wave, even if it has a low frequency (see Table 10). Otherwise, without oversampling, the sampling frequency is twice the signal frequency. In an extreme case, two PWM periods on/off are sufficient to generate a sine wave. Ideally, however, you would need a steep filter at twice the PWM frequency.

Resolution N [bit]	2^N	Voltage at 10 V Full Scale	ppm Full Scale	% Full Scale	dB Full Scale
8	256	39.1 mV	3'906	0.39	-48
10	1'024	9.77 mV	977	0.098	-60
12	4'096	2.44 mV	244	0.024	-72
14	16'384	610 µV	61	0.0061	-84
24	16'777'216	596 nV*	0.06	0.000006	-144

* 600 nV is the 25°C thermal noise for 10 kHz bandwidth of a 2.2 kΩ resistor.

Table 9. Resolution and least significant byte at the ADC.

- It is possible that insufficiently suppressed PWM clocks are lurking around in the system.

Notes on the design of a DAC using PWM:

- Since it takes some clock cycles to load the new target value into the counter, you cannot load too small target values, otherwise the timer is already at the target value when you load it. The actual resolution is always smaller than the counter bit width.

- For a pure alarm or signal tone, use the PWM frequency itself as the audio frequency, with a 50% duty cycle. Sounds not too bad if the highest frequencies are attenuated a bit, e.g. by an RC low-pass filter with a cut-off frequency of 8 kHz.

Maximum Signal Frequency	Oversampling Factor	PWM Frequency	Resolution = Counter Bit Width	Clock Frequency
$f_{signal,max}$	m	$f_{signal,\,max} \times 2 \times m$	n	$f_{PWM} \times 2^n$
3.13 kHz*	16	100 kHz	8	25.6 MHz

Attenuation of an RC low-pass filter with a cut-off frequency of 3.13 kHz at 100 kHz: 30 dB.

* Minimum bandwidth for voice, formerly used in analog phones.

Table 10. Example of frequencies for a PWM-based DAC.

30 Component Selection

- For sine signals, a table (i.e. memory space) is required for at least a quarter of the sine, since the sine calculation is usually not possible within the time for resetting the counter. Different frequencies are achieved by reading this table with different step widths, i.e. at the highest frequency the table is read value by value with the PWM frequency. At lower frequencies, the PWM counter is loaded several times with the same value from the table until the next one follows.

For $1 per 1000 pieces, you can get the DAC081S101 from Texas Instruments, a very well specified DAC with 1% accuracy over the whole temperature range and a maximum frequency of 30 MHz, controllable via SPI. This is a clear quality improvement compared to most microcontroller DACs and a resolution usually unattainable with PWM-based DACs. Of course, this comes at a price.

Figure 12 shows a well-known discrete DAC circuit. This solution has two advantages: it is maximally fast, as fast as the microcontroller's ports can be set, it is inexpensive from a component point of view, and only resistors are needed. These advantages can be tempting to accept parallel control, which requires a number of microcontroller pins dedicated to it.

However, the circuit has three serious drawbacks. First, the accuracy of the analog voltage depends on the tolerance of the resistors. The correct tolerance, i.e. the tolerance at operating temperature and at the end of life, must be used. With 1% resistors, whose end-of-life tolerance is often 2–3%, you are already at 1.4% tolerance with the 4-bit resolution shown in Figure 12.

Second, the analog voltage depends on the accuracy of the port voltage. Even if you use a buffer such as a 74HCxx device, the problem remains. Since the output voltage of the digital device depends on the supply voltage, its variations directly affect the output voltage.

Finally, you must perform a significant amount of testing to make a reliable accuracy statement. This last reason is often the main obstacle to implementing discrete circuits. The development and test costs of a discrete circuit compared to the cost of an available integrated circuit are only justifiable in very high volumes.

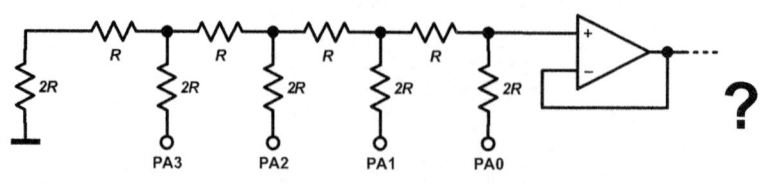

Figure 12. Fast DAC with R-2R network. PA0 to PA3 are the GPIOs of a microcontroller. The question mark indicates that the circuit should be used with care according to the notes in the text.

Microcontroller Selection

The most complex IC in a circuit is usually a microcontroller or an FPGA. Unfortunately, these types of ICs usually have to be defined at the very beginning of development. Changing the type later is like open-heart surgery. Therefore, good advice is needed when choosing these highly integrated devices, and it is given here as follows: Choose a microcontroller or FPGA family that has been used successfully in a previous design. If this is not possible, it must be clear to all project participants that the first project with a new microcontroller or FPGA family usually requires a development effort that can only partially be passed on to the customer.

The example in Table 11 illustrates the above statement. Two companies known to the author followed two different microcontroller selection strategies. One year after the start of the companies, the balance could be drawn according to Table 12.

Do not underestimate the effort required to use a previously unknown microcontroller. It includes the following:

- The programming environment is new, the toolchain needs to be set up and learned, compiler options need to be set correctly.

- Certain code libraries may not be available, not portable, even if written by yourself, expect rewriting of your own or foreign code.

- There is no existing working setup of the microcontroller, i.e. a pattern of complete registers correctly set for input/output pins and for the interrupts.

	Strategy from the beginning of the company, five employees.
Company A	Evaluate the "best" microcontroller type for each project.
Company B	Microcontroller projects with "Blackfin" types only (combination of a 32-bit RISC processor and a 16-bit fixed-point DSP from Analog Devices).

Table 11. Two microcontroller selection strategies.

32 Component Selection

	Balance sheet, 1 year after starting the business.
Company A	• Five microcontroller projects. • Five different microcontrollers from different vendors used. • Deadlines and costs exceeded in most cases. • Poor quality of execution. • Three projects in the red. • Dissatisfied customers. • Employees dissatisfied. • Know-how fragmented among employees.
Company B	• Five microcontroller projects with a dedicated 32-bit family. • Innovation award with a customer. • All deadlines and costs met, in some cases even undercut. • Quality of execution according to requirements. • Company was able to create some reserves. • Customers satisfied. • Employees satisfied. • Know-how well distributed among all employees.

Table 12. Results of different microcontroller selection strategies.

- Hardware and software tools for using a specific microcontroller may be missing. For projects with large code size, an in-circuit emulator (ICE) should be used, despite the powerful in-circuit debugging capabilities. Time-critical sequences can be analyzed better, if at all, with an ICE. Also, the breakpoint possibilities—complexity and what can be reacted to—are more extensive. Experience shows that the higher cost of the ICE is worth it for a complex project because of the time saved in debugging.

- There may be no incentive to create program code for reuse because the next project may involve a different microcontroller type. This can lead to unclean, "quick and dirty" programming.

- Due to lack of experience, it is difficult to estimate whether the addressable memory and performance will be sufficient.

For low-volume products, it may make sense to use an obviously overpowered microcontroller in a new project simply because you can copy and paste almost everything from a previous project.

For problems with heavy signal processing, i.e. filtering, convolution, etc., but low data throughput, your designs will be more expensive but safer with a microcontroller with a floating point unit (FPU). Mastering filter calculations on a non-FPU microcontroller can easily cost you two weeks without prior experience. In addition, numerical accuracy, no overflow, and stability must be rechecked with each change to the signal processing code.

A commonly cited reason for choosing an Advanced RISC Machines (ARM) Cortex-M based microcontroller is the ability to port code. This may be to a smaller or larger core, to a Cortex-M microcontroller from another manufacturer, or you may want to upgrade to a more powerful Cortex-M type. Keep in mind that a Cortex-M class is an intellectual property platform on which the licensee builds its own dedicated hardware. Certain parts are strictly defined, such as the Cortex microcontroller software interface standard, the instruction set, and the structure of the core. However, the peripherals can be variable and are not guaranteed to exist [R150]:

- Floating point unit (FPU) on Cortex-M4 and M7.
- The wake-up interrupt controller.
- The memory protection unit.
- The micro-trace buffer.
- A certain number of interrupt levels; these can vary from 8 to 256.
- Any peripheral blocks do not belong to ARM licensing.

Consequently, even if the core remains the same, the code must be rewritten if it is to be used on another Cortex-M type. Usually, the peripheral module code is the part that requires the most work, and it is precisely this part that must be ported and cannot be transferred to another Cortex-M device.

More than one Microcontroller?

Packing all the functionality into a single MCU is attractive from a cost perspective. However, the more complex the software becomes, the more difficult it is to debug. Using multiple, identical, smaller microcontrollers has several advantages:

- The software on the individual microcontrollers is leaner.
- The software can be developed independently for each MCU.
- Each microcontroller can be tested with its own environment.

34   Component Selection

- The risk of side effects during firmware updates is reduced.

- Troubleshooting is simplified.

Do not attempt to use multiple microcontrollers in time-critical and/or safety systems. Implementing a task scheduler across the microcontrollers would be challenging. Instead, use a single microcontroller with a real-time operating system (RTOS). This advice is based on an experience I once had. For an advanced battery-powered drill that automatically stops at the desired hole depth, a second microcontroller was added because the computing capacity of the first single microcontroller was overwhelmed. However, this created new problems because time-critical communication was not working properly. Eventually, we moved to a single, larger microcontroller. Weeks of development time and money were wasted.

Safety Microcontrollers

Safety microcontrollers are a special group of microcontrollers. They are characterized by the following features:

- Dual or triple microcontroller cores that operate in lockstep, processing the same program a few cycles apart and as far apart as possible in the microcontroller package.

- Built-in hardware error correction code for flash memory and random access memory (RAM) accesses, detecting and correcting memory and bus errors.

- Built-in CPU and RAM self-test.

- Peripheral modules are also at least parallel.

- Built-in error response module to facilitate error handling.

Manufacturers of safety microcontrollers typically provide not only components, but also guidance on how to use them, up to and including decoupling for safe programming.

Despite their impressive safety features, safety microcontrollers are only useful when significant external interference is expected. In general, the risk of malfunction due to a programming error is much higher than due to an external influence. A safety microcontroller does not protect against poor quality firmware. Therefore, if you are free to use a safety microcontroller, consider whether the additional cost of implementing a safety microcontroller is better spent on more extensive firmware testing.

If you need the highest computing power for conventional programming without the use of an FPGA, you can only get it with a processor system. For tasks with high computing power, the question of whether to use a processor system may arise. In this case, the following must be considered:

- Designing your own motherboard today is both pointless due to off-the-shelf availability and virtually impossible for non-specialists due to complexity.

- A processor has a shelf life of about two years (i.e., a successor is released every two years), which means that a processor can quickly become unavailable.

- Similarly, a motherboard that is available today can be expected to be unavailable in two years. Sure, there will be a successor product; it will probably be compatible, but you will have to run all the system tests again for safety and standards.

Consider splitting the solution into a piece of electronics that you design yourself and a laptop or tablet from a third-party manufacturer.

It is safe to say that all electrical engineers are familiar with one or more microcontroller families. The same cannot be said for field-programmable gate arrays (FPGAs). However, there are cases where an FPGA is technically a better solution than a microcontroller, albeit at a higher component price. Real-time processing with more than one time-critical input signal, for example, is only safely possible with an FPGA due to its multi-parallel architecture instead of a conventional microcontroller that works sequentially with one core. Strictly speaking, however, this is only true for FPGA utilization levels of up to about 50%. Beyond that, existing structures are recompiled and thus optimized, which can result in different propagation delays for the same signals. On the other hand, any addition to the microcontroller automatically causes a change in the sequential flow, with the potential for delayed processing of a time-critical signal. If you know at the beginning of the project that you will be processing many real-time signals, the only argument against an FPGA is price.

Microcontroller GPIO Pin Notes

Set any unused I/O pin as an output and provide it with a test point; you may need the pin later. If you want to be sure or do not trust the firmware, set a pull-up resistor. Normally, the microcontroller I/Os are wired as high impedance inputs at startup. Interference can cause an input stage with an undefined input potential to switch permanently and dissipate power unnecessarily. However, a microcontroller does not fail because of unconnected pins; it simply has a higher power requirement under certain circumstances.

Provide all hardware configuration pins with 10 kΩ pull-up resistor and 200 Ω pull-down resistors unless otherwise specified in the datasheet. The pull-up resistor must not be too large, otherwise it will result in a logically undefined state together with the input resistor.

Do not connect a GPIO microcontroller pin directly to VCC or ground. Use pull-up resistors and pull-down resistors. During production testing, over-voltages on VCC and negative voltages on ground may occur. The power pins can handle these variations, but the GPIO pins may not. All interrupt inputs must be kept in a passive inactive state with an external pull-down resistor or pull-up resistor.

Microcontroller Clocking Notes

Should I use the microcontroller's internal clock generator, a quartz crystal, or a ceramic resonator? The internal clock generator is the cheapest and requires no additional elements. However, it is far too inaccurate to keep time, and often interfaces also require a low tolerance clock—check the specifications. In this case, a quartz is the usual choice. It also maintains its frequency well over a wide temperature range. On the other hand, a quartz is a fragile part. It can break if subjected to shock, and it is not particularly ESD resistant. Ceramic resonators are something in between the internal clock generator and the quartz. Their accuracy is about 0.5% or 500 ppm compared to 10–30 ppm for a quartz. In contrast, ceramic resonators are shock resistant and tolerate ESD events well. This may be a reason to use them, but if not, a crystal may very well be cheaper for the same frequency, i.e. you get a much more accurate device of the same size for less money.

A question that is often asked is: Should the crystal case be grounded or not? I always solder it to ground. I don't see any disadvantages. The load capacity may need to be adjusted, but the grounded case provides shielding, and soldering the case provides additional mechanical stabilization.

Microcontroller External Memory Notes

If you need to expand the microcontroller's memory with external memory, the following tips may help:

- The address space should be defined early in the design phase.

- All direct memory access (DMA) addresses and control pins must be kept in a passive, inactive state with external pull-down resistor/pull-up resistor.

- By setting pull-up resistors and pull-down resistors on the data bus, a defined state exists when non-existent memory is addressed.

- Provide a 22 pF capacitor directly at the DMA request pin of the microcontroller and ground; do not populate. This may help with EMC problems.

- DMA accesses that reach the board from the outside must be provided with overvoltage protection.

For cost reasons, you will probably want to use the smallest possible memory devices for the finished product. However, you should provide the layout and circuitry so that the next larger memory device can be assembled.

If an external Flash memory is used for the boot code, choose one that is at least twice the size needed to accommodate two pieces of firmware. During the power-on reset, the first part performs a cyclic redundancy check of the second main part. Only if the test confirms the data integrity, it continues with the main part.

If dynamic random access memory (DRAM) is used for data and/or the stack, check how it works with the debugging tools. It may be necessary to add a control input to the memory controller that allows the memory to be refreshed while preventing any reading and writing to the DRAM when using an ICE.

For high reliability products, store the data in an error-detecting or even error-correcting format. Also, provide additional circuitry for writing to memory, so that more than one control line must be set correctly for writing. If the program gets out of control, it is unlikely to overwrite the data in memory.

Prefer RAM with an active-high chip select, and design the system so that chip select and the write control line are inactive when the system shuts down. In addition, active-high chip selects on the RAM and active-low chip selects on the rest of the peripherals can be used to strictly separate these two areas.

If you are using RAMs with multiple chip selects, either connect them as shown in the reference design, or make provisions to return to this configuration by changing resistors.

Microcontroller Reset Notes

The basics are:

- Do not rely on the MCU's internal reset pull-up resistor, add one externally.

- Add a 22 pF capacitor directly to the microcontroller's reset pin and ground.

- Add a reset button even if space is tight, you will need it often during testing.

- Keep the circuit around the reset button clean from fast signals and design it for safety: decoupling capacitors, avoid crosstalk possibilities and keep PCB traces short.

- Some microcontrollers require thousands of clock cycles with active resets to fully initialize their internal logic, check the datasheet. Make sure the reset is active for a long enough time. In the case of a reset button, this can be achieved with a debounce circuit (see above). This saves a lot of troubleshooting later.

- If the microcontroller receives the reset from another device, consider disabling this connection for testing purposes, e.g. with a jumper or a 0 Ω resistor bridge.

Why You Should Use an External Watchdog

Microcontrollers usually contain an internal watchdog, but do not use it for safety reasons. In the event of a massive disturbance, such as an electrostatic discharge (ESD) event, the watchdog register may be altered, disabling the watchdog. Use a dedicated, non-programmable watchdog device, which is considered more immune to ESD.

If a watchdog is used, it is highly recommended that a jumper be placed between the watchdog and the microcontroller in the prototype for quick disabling. Do not connect the watchdog output (WDO) directly to the reset input (RESET) of the microcontroller. If you have not already done so, the following points need to be clarified:

- Microcontrollers can have very specific requirements for the edge steepness and duration of reset signals. Is the WDO signal compatible with RESET?

- The watchdog reset input (WDI) must be AC coupled or edge-triggered so that a change of state is required to reset the watchdog.

- Is the watchdog inactive for a specified time after a RESET or power-up to allow the system to initialize itself? Is the selected timeout long enough?

- If the watchdog has not been reset since the last watchdog reset, does the WDO trigger a RESET again after a certain time?

Other notes:

- The WDO must trigger a non-maskable MCU reset input.

- Typical timeout values are 10 ms to 2 s, depending on how long the system is allowed to go out of control and become dangerous.

- On the other hand, the time-out must be long enough so that WDO does not trigger a RESET during normal operation and high processor load.

- WDI must be reset by the main program, not by an interrupt routine, because the latter can continue to run despite a stuck main program. Perform the WDI reset from exactly one line of code, usually somewhere in the main loop.

- The watchdog should not be clocked by a system clock, but should have its own clock, if it is needed at all.

If the watchdog has a manual reset input and a reset output other than WDI, connect both to normal microcontroller pins. This way you can determine if the watchdog is working. Strictly speaking, you cannot check the watchdog timer, but you can check if the watchdog is powered at all. Of course you have to use a separate circuit to debounce the manual reset button.

Notes on Low Level Programming and HAL

The concept of the hardware abstraction layer is promising: It allows maximum portability of code. Table 13 looks at some other aspects and draws a comparison with low-level coding.

It is tempting to use HAL because you can start even if the microcontroller type is not yet defined. Also, you code fast and rely on readily available structures—there is no need to study registers. But this is

Criteria	**Low Level Code**	**HAL Code**
Code size	Small	Large
Speed	Fast	Medium
Portability	Medium	Good
Readability	Medium	Good
Functionality	Full	Certain features may not be available.

Table 13. Comparison of low-level code and HAL-based code.

only true at the beginning or for very small projects. As soon as the code becomes more complex, problems arise. For example, the STM32 HAL has a stateful middle layer that seems to be used to support multi-drivers and multithreading. But what it does is block resources that should never be blocked. When you write true multithreaded code, the HAL blocks processing and the real-time latency of UARTs, SPIs, and I2Cs is serialized. No wonder your code stops working! I recommend using low-level code whenever possible. There have been too many times when I have had to rewrite HAL code because it just did not work.

Notes on Debugging Microcontroller Firmware

From a hardware perspective, and without an in-circuit emulator (ICE), there are two ways to debug:

- Direct connection between the integrated development environment (IDE) and the microcontroller, usually using a Universal Serial Bus (USB) to Joint Test Action Group (JTAG) converter. A common converter IC is the FT2232H, which is usually placed directly on the target board. This is a slow connection, and due to its low bandwidth, tracing (i.e., recording the progress of the code in real time) is not possible. Hardware breakpoints are limited to those provided by the microcontroller. There are unlimited software breakpoints, but they must be embedded in the code and, if changed, require a new build and download.

- Indirect connection between the IDE and the microcontroller using an intelligent debugging IC with its own memory. For example, the Segger J-Link provides tracing and an unlimited number of hardware breakpoints.

From a professional point of view, the worst setup for debugging is using a USB-to-JTAG converter and downloading the code only to the microcontroller's Flash memory, as used in the Arduino environment. It is painfully slow, and there is no way to directly control what the microcontroller is doing. With an intelligent debugger, you can step through the code or record a sequence that needs to be run in real time for later analysis.

Programming the Microcontroller in Production

The following basic variants can load the firmware into the flash memory of a microcontroller:

- The manufacturer or distributor programs the microcontrollers; they are only soldered in production. About 50% of microcontrollers are programmed this way [R171]. Of course, it must be crystal clear which revision of the firmware is to be programmed.

- The other option is in-system programming. The microcontroller is programmed during production using a JTAG interface. A dedicated connector is usually only justifiable in terms of price if later firmware updates are possible. Otherwise, an edge connector may be an alternative, or programming via an in-circuit tester, i.e. a bed-of-nails tester. Of course, this increases test time and may not be feasible for large lots.

Choosing Field-Programmable Gate Arrays

For many electrical engineers, FPGAs are less well known than microcontrollers, so let us have some application notes:

- Static random access memory (SRAM)-based FPGAs are typically a step ahead in performance. However, they have a slow start-up time, typically more than 100 ms.

- Flash-based FPGAs have the advantage of immediate startup with possible code changes.

- One-time programmable (OPT), anti-fuse based FPGAs start immediately, are copy protected, more energy efficient, require no external memory, and are radiation resistant for satellite use.

- The high operating current of FPGAs can be a problem for mobile devices, but there are low-power FPGAs available, such as those from QuickLogic.

- FPGAs are relatively easy to convert to an application-specific IC (ASIC). For very high volumes, ASICs are mandatory because of the low price per IC.

As with microcontrollers, FPGAs can only be used profitably if the code is reused in multiple projects. Therefore, the selection of an FPGA family is determined by the previously used FPGA types and the available, well-established tool chain with its own code library.

Usually, except for very computationally intensive algorithms, the limiting factor is not the possible logic complexity, but the number of digital inputs and outputs (I/Os). Therefore, I list the required I/O pins first and then the required number of logic elements (LE), which is equal to the number of logic cells. A logic cell consists of a lookup table and a register plus associated logic. Higher-level units such as "configurable logic blocks" or "logic array blocks" are not standardized terms across vendors. Finally, I consider the number of clock domains.

For speed, the rule of thumb is that a higher speed level costs about 20–30% more for a 12–15% increase in performance. For a product with a lot size of more than 100, the speed level should be kept as low as possible by very efficient programming, otherwise the use of an FPGA may be too expensive.

FPGAs with an integrated hardware microcontroller are available ("hard core"), or a part of the FPGA can be programmed to emulate a microcontroller ("soft core"). ARM says about the latter: "With a size of ~150',000 LE, the Cyclone V FPGA is large enough for the implementation of most Cortex-M systems". All of these solutions are useful when both a microcontroller and an FPGA are needed in the system, (i.e., where an FPGA must be used, anyway). The advantages of integrating the microcontroller into an FPGA are as follows:

- Less space is required.

- Direct connection of the FPGA to the microcontroller, resulting in less external interference and a faster connection.

- Computing can be shared between the MCU and the FPGA.

- Interfaces can be upgraded to the latest standard without hardware modification because they are implemented in the FPGA.

- The debug interface to the microcontroller can be freely designed.

- The microcontroller in the FPGA runs like an emulated microcontroller, simplifying monitoring and debugging.

- No additional crystals for the microcontroller.

The disadvantage is that the traditional division of work between the microcontroller programmer and the FPGA programmer can no longer be well implemented. If more than two programmers are planned to be involved, have a thorough discussion about whether the team is willing to work closely together.

Notes on FPGA Reset and FPGA Clocking

Route the reset signal to an input that you have Schmitt-triggered if the reset signal has a slow edge, such as after a de-bounce circuit. Note that the FPGA may need thousands of clock cycles to fully boot. A watchdog must not trigger a reset before then.

Clock ICs are expensive and may tempt you to build your own clock circuit. Unfortunately, designing a clock circuit is fraught with pitfalls:

- Too high an operating power level will cause frequency instability and possible damage to the resonant element. If the operating level is too low, the oscillator will start up slowly, possibly within seconds, or not at all, and will be sensitive to interference.

- Special techniques like spread spectrum can hardly be realized with a reasonable discrete circuit.

- The circuit must be temperature tested, possibly age tested, which can be time consuming.

- The layout of the circuit is critical. It must be ensured that no logic lines run near or through the normally high-impedance oscillator circuit, as they will copy into it and cause frequency instability or jitter. The stray capacitance of the layout can easily be up to 10 pF and must be taken into account.

- Always measure the clock signal on the first prototype and compare it to the conditions set by the FPGA; this will save you a lot of trouble later if it turns out that the clock is not clean or correct.

After these explanations, designing a separate clock circuit is not economical for small quantities.

A clock module is more space efficient than a discrete clock circuit, which in itself can be a sufficient reason to use one instead of a discrete clock circuit.

Use spread-spectrum clock modules if no specific frequency is required; this makes it easier to pass EMI tests [R009].

Should you choose a crystal-based or MEMS-based module? Table 14 compares the two types and draws conclusions.

It should be noted that there are no standard packages for clock modules, so you need to pay close attention to availability, although this is not so problematic for small quantities.

Clock modules can have a high drive current (e.g. 40 mA). This can destroy components that are incorrectly connected to the clock generator, e.g. if the clock receive pin is incorrectly set as an output and connected to ground. You may want to place a resistor between the clock module and any clock receiving pin. This not only limits the current, but also reduces the clock spectrum, since the resistor and the input capacitance of the FPGA pin form a low-pass filter. However, do not

Criteria	Quarz-based	MEMS-based
Stability	<1 ppm, with temperature controlled (OCXO) down to ppb	
Phase noise, jitter	Very low	Moderate, susceptible to micro frequency jumps
Current consumption	Very low	Medium, may result in longer startup times
Vibration	May break	Robust
Mass production time	Medium	Low
Cost	High	Low
Radiation resistance	High	Low
Example application fields	Precision circuits, RF- and microwave circuits, RADAR, space	Automotive

Table 14. Comparison of crystal-based vs. MEMS-based clock modules.

set the resistor value too high, otherwise you will need a Schmitt trigger circuit at the clock input, because the clock signal will stay in the logically undefined area for too long.

Check if you can still see some of the clock signal at the supply pin of the clock module. Adjust the decoupling capacitors accordingly.

System-on-Chip Selection

Following the discussion of the fusion of an FPGA and a microcontroller, a programmable system-on-chip (PSoC) must be mentioned. These are a collection of predefined blocks, such as a microcontroller core, logic blocks, peripheral blocks such as an I²C interface, an ADC, operational amplifiers, encryption modules, graphics processors, and even radio modems. PSoCs are the most widely used in mobile phones. But this is also a warning: The world of PSoCs is fast-moving, driven by the need for maximum integration into commercial devices. Portability between PSoCs from different vendors is poor. As a result, PSoCs are only recommended for products that will be upgraded to a new version one or two years after the current version goes into production.

Integrated Circuit Package Considerations

ICs come in different packages. Which packages should be preferred? There are three basic issues to consider: pin accessibility, post-solderability, and lead inductance.

It is not necessarily wise for a prototype to choose an IC in a package where the pins are not accessible after soldering. This is the case for ball-grid array (BGA), dual-flat-no-lead (DFN), quad-flat-no-lead (QFN), and very thin leadless array (VTLA) packages. Reasons for doing so are space constraints, or if it must be this specific IC and it is not available in another package. Otherwise, it is clearly an advantage for the designer to have access to any IC pin for debugging purposes.

As for re-solderability, if a broken IC cannot be replaced in the prototype, the whole board is potentially unusable, and you run out of prototype boards. Now, according to Murphy's Law, it is that one non-re-solderable IC that will fail.

Experience from people who have tried has shown that a BGA package with many solder points cannot be soldered by hand. The only proper way to solder a multi-pin BGA IC is to use a soldering oven. This oven uses a special thermal profile to ensure that the PCB and BGA IC are heated evenly without creating mechanical stress.

BGA ICs with multiple rows of pins are difficult to remove from the PCB without damaging one or the other, or both. PCBs with defective BGA components usually have to be discarded. There are BGA adapters that allow the IC to be removed, but experience has shown them to be unreliable in terms of contact, and they are not recommended here.

The QFN package can be soldered by hand if the thermal pad does not need to be soldered. If soldering is required, a soldering oven is required. It is also possible to work with a heating pad, but then the microcontroller has to be soldered first. In practice, it is not possible to desolder a QFN type whose thermal pad is soldered to the PCB.

The situation is different with thin-shrink small-outline package (TSSOP), quad flat package (QFP), and the like, i.e., those that have a very high pin density but whose pins are accessible from the side and do not contain thermal pads. If there is a solder mask between the pins, I use a lot of flux to pull the tin to the right places. Alternatively, or if there is no solder mask between the pins, I use a lot of tin, which initially creates rough shorts between the pins. I remove these with solder braids. In both cases, you do not need a very fine soldering tip; in fact, such a tip is more of a hindrance because it can only transfer a small amount of heat.

When fast signals are present on the board, low trace impedance is required for high-speed components. The available data or measurements in the datasheet are important for this. Lead inductance varies widely from less than 1 nH to 15 nH [R121]. The BGA packages have the lowest lead inductance. If you do not keep this in mind, you may have to change to another package with a redesign.

If you are forced to use a BGA package, be aware of the following:

- BGA connectors are not mechanically stable. Contacts can break if the PCB is bent or shaken. Also, contacts can break when the PCB heats up because of the different coefficients of expansion of the PCB and the BGA IC. For this reason, plastic BGAs are better than ceramic BGAs. One solution is to glue the IC to the PCB at the same time as it is soldered in the oven.

- Whiskers can short out pins. Whiskers are crystals that can grow out of lead-free solder. This must be prevented, especially in safety-critical applications, which is why leaded solder is permitted in such applications. Whiskers are also prevented if the BGA IC is glued during soldering. If enough glue is used, the space between the pins is filled with glue.

- With a pitch of 1.27 mm and a maximum of about 400 contacts, 40% of which are ground or supply pins, the BGA IC can still be placed on a conventional PCB without connection problems. Dog bones and ground floods are possible, and two 125 µm wide traces can be routed between two pins.

Pitches of 1mm and below require the use of High Density Interconnect (HDI) boards. HDI is high technology in terms of manufacturing requirements, but it is state of the art today, as almost all mobile consumer devices are made from such boards. HDI PCBs do not use glass fiber epoxy material (i.e., no FR4) because the holes are shot with a laser, which is not possible with a material containing glass. Using a laser, it is possible to make tiny vias with diameters less than 0.2 mm. An HDI board is built layer by layer like a conventional board, but the holes in each layer are lasered before assembly, allowing blind vias and especially buried vias in the inner layers. This makes it possible to unbundle dense BGA IC connections on different layers. These vias also have better specified electrical properties than FR4 vias. Impedance-controlled traces can be created despite small layer spacing because very thin traces (e.g., 100 µm (4 mil) wide) can be created. Peripheral Component Interconnect (PCI) Express buses or Double Data Rate (DDR) memory access can only be implemented with HDI boards.

The design of an HDI board must always be discussed in advance with the manufacturer. Not everything that can be drawn in the layout program can be manufactured. In addition, many parameters are interdependent. For example, the microvia diameter of a blind hole depends on the thickness of the outermost layer. Through-plated vias through all layers are expensive on the HDI board. In general, an HDI board is more expensive than a conventional board.

Component Selection and Obsolescence

One of the electronics designer's biggest concerns is: "What if a particular component is suddenly out of stock?" Large companies often employ at least one "component engineer" to prevent this from happening, among other things. Component engineers evaluate components based on the design engineers' specifications and maintain a database. They create a "preferred component list" of items that have a particularly favorable relationship to price, performance, and deliverability. Component engineers are particularly concerned with ensuring component availability over the defined service life. They look for a second source. Sometimes they even list pin-compatible alternatives. Preferred components are usually less expensive because they are used for a large number of projects. In my experience, such a service is very helpful in development.

If your company does not have a dedicated component engineer, you will have to perform this function yourself, which can be difficult due to lack of time. The simplest way to mitigate this problem would be to purchase enough of the selected components to use them in all the devices you have ever made. Unfortunately, in many situations, you are either not allowed to make this "lifetime purchase," or you simply do not know the final number of units, or both. However, there are situations where the exact number of units to be produced is known. One example is the production of a certain type of car, where the number is usually known before production begins.

If a lifetime buy is not an option, here are some tips to avoid the out-of-stock problem.

Jack-of-all-trades ICs are dangerous; for example, an IC that combines a watchdog function, a manual reset input, a voltage threshold detector, and clean reset generation during startup, shutdown, and brown-out. Such an IC is advertised as "reducing circuit complexity," which is true. However, there is no pin-compatible alternative. If it goes out of stock, a layout redesign is inevitable. Imagine telling your boss that a redesign is necessary because the watchdog is no longer available! So stay away from jack-of-all-trades ICs without good reason. For separate common functions, use separate common components. For a watchdog function alone, I can use the TPS3823 from Texas Instruments or the ADM6823 from Analog Devices interchangeably; they are pin and function compatible.

If a component is only available in the automotive temperature range of -40 to 125°C, but you are not in the automotive industry, consider the following: As mentioned earlier, the automotive industry typically buys large batches of components. If you need a large lot, you may end up competing with the car manufacturers and losing. It is easy to fall into this trap because there are many automotive-only components (see Table 15).

Application	Temperature range	Proportion of op-amps at Digi-Key
Commercial	0°C to 70°C	18%
Industrial	-40°C to 85°C	37%
Automobile	-40°C to 125°C	37%

Table 15. The main temperature ranges of ICs and their share of the op-amps in the distributor, Digi-Key.

Even worse, a component may only be available in the commercial temperature range of 0°C to 70°C. These are often components for mass-market devices such as USB sticks. Device manufacturers buy "lifetime" because the next device is already in development and the old one will not be repaired. Components from this range are usually discontinued quickly because the pressure to improve is immense and the manufacturer cannot get rid of the "obsolete" components.

For good availability, it is therefore best to use an IC that is (also) available in the industrial temperature range. Due to the long lifetime of industrial equipment, there is a constant demand for such components over a period of years, and the manufacturer of the component adapts to this with an appropriate follow-up production plan.

In a positive sense, the discontinuity problem can be countered with standard components. A standard component can be used in large numbers in different market segments, namely commercial, automotive, industrial and military. Even if the original manufacturer goes out of business, smaller companies emerge to continue manufacturing the component under license. The best example of this is the 8051 processor introduced by Intel in 1981. This processor is still in full production today, now at Silicon Labs. Use the following indicators to identify a standard component:

- Various distributors stock large quantities of the exact component—use the full component name.

- Available from various distributors as whole reels and original reels, not re-reels.

- Available from distributors, not manufacturers, in a variety of package styles and temperature ranges.

- The component has been around for several years, as indicated by the date on the datasheet.

- The component is one of the least expensive of its kind.

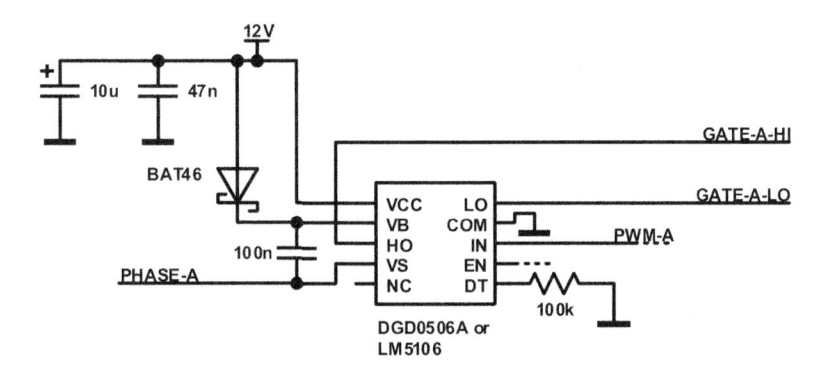

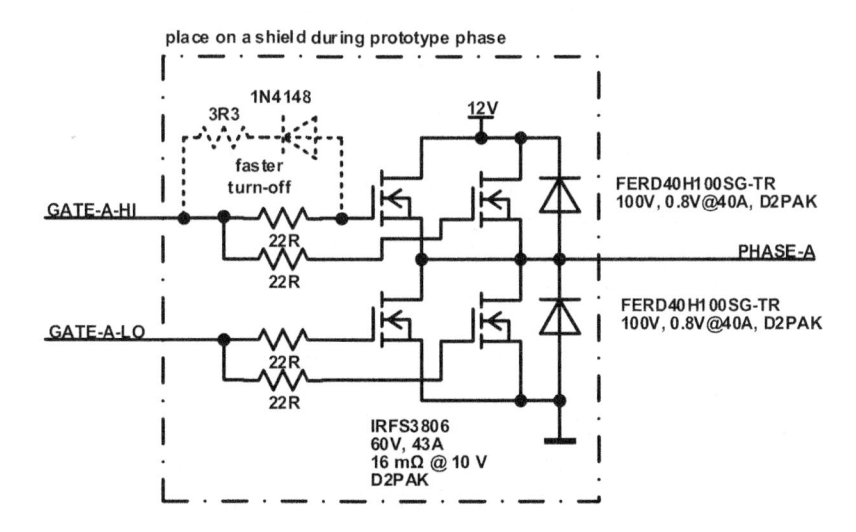

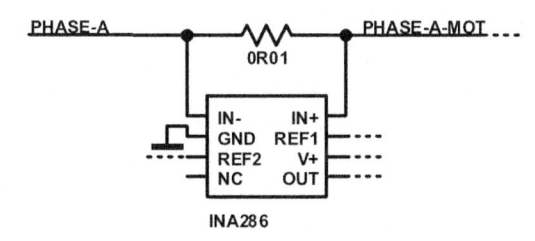

Figure 13. Example of a discrete motor control circuit.

- The manufacturer's standard lead time is short, for example, six weeks or maybe three months. A long lead time, such as one year, indicates infrequent production.

- Explicit mention in literature and application notes from manufacturers other than the component manufacturer.

Rather than scouring the Internet for each distributor's site or each manufacturer's site, I often use aggregators such as octopart.com, findchips.com, or ciiva.com.

As an example of the discussion of using standard components, Figure 13 shows a discrete motor driver structure. All of the semiconductors and ICs can be easily replaced with functionally and pin-compatible alternatives. However, the circuit does not implement overcurrent protection, thermal shutdown, or gate charge monitoring. A motor driver IC would provide all this, but you would be completely dependent on that specific driver IC type. A middle ground is to choose a more feature-rich gate driver, such as one that includes current monitoring, but still allows free selection of transistors.

In recent years, there has been a clear trend against the use of standard components. The reasons are as follows:

- The product lifecycle has become very short in some cases. The aspect of long-term availability is no longer relevant. The product is produced for a short time, much shorter than the time a component is usually on the market.

- The product is at the edge of feasibility; the latest elements must be used to achieve the desired performance. Perhaps the performance could be achieved with several standard components, but the price/performance ratio would be worse, especially if more space is needed.

- Standard components are mediocre in all parameters, which results in sub-optimal solutions that do not stand out from the competition. Often the customer is willing to pay a little more for a better-quality product.

- Changing the manufacturer of a standard component can cause problems, even though the element is basically "the same". I had a case where a BSS123 transistor from one manufacturer worked in the circuit; the "same transistor" from another manufacturer caused problems. Standard components, for example, can tempt a buyer to change the manufacturer of a component at short notice, e.g. in the case of a currently more favorable price offer, because the change seems to be technically irrelevant. Proprietary components (i.e. components produced by a single manufacturer) are usually much more fully specified.

- Manufacturers have good search capabilities on their product pages; it is quite easy to quickly narrow down the selection with a few specification values and find the most suitable component.

Potential Problems with the Assembler

The major component distributors offer "re-reeling" services, i.e., rewinding cut tape onto a reel. A leader and trailer are added according to the Electronic Industries Alliance (EIA) standard. However, an assembler may sometimes need to add a leader strip to a rewound reel because his machine is not running smoothly with the supplied leader strip. Assemblers are used to reeling; it has always been a part of their service. The big distributors have taken some of this service and revenue away from the assemblers without any benefit to the customer. The assembler usually wants to charge for a re-reel, and this leads to disagreements with the customer, who takes the position that the re-reels are produced according to standards. This can be avoided by providing the assembler with only original reels or tape sections.

Assemblers expect a certain reserve in the delivered material, usually 10% to 20%. For the most expensive component, you may want to go below that, but even there, a handful more than necessary is advisable. Every now and then, for example, there are breakdowns; the assembler must switch to another machine, and some components are lost from the strip with which she or he started. There is nothing more troublesome than when production comes to a standstill because a component came up short!

Sensor Selection

Sensors mediate between other physical quantities and electrical currents or voltages. Electrical engineers are not physicists and can easily be misled by inadequate information in data sheets without realizing it. It is thus advisable to design the system so that at least two sensor types from two different manufacturers can be connected.

More and more sensors are available in digital versions. If your product is to be a precision device and a digital sensor IC is available, you can forget about a solution with an analog sensor and an analog signal processing circuit. For example, the multicomp HCZ-D5 capacitive humidity sensor, which consists only of a capacitor with a special polymer dielectric, has a tolerance of ±5% in the capacitance-to-humidity ratio. It is very inexpensive, but you need a fairly extensive circuit to accurately measure picofarad deviations. Most likely, this circuit will have a tolerance of at least 1% over the entire temperature range. In comparison, the IC SHT21 sensor directly provides a digital humidity value with a guaranteed tolerance of ±3%. The latter is possible because all transistors in the SHT21 are fabricated on the same die, resulting in closely matched parameters. In addition, the SHT21 is calibrated after production and the calibration data is stored on a chip.

Display Selection

The classic custom LCD still offers the following major advantages over a pixel screen:

- Excellent power savings.

- No graphic routines that burden the microcontroller.

- No graphics memory required.

- Custom LCDs are cheaper than a graphic display from a lot of about 1000.

Color effects can be achieved by overprinting LCDs with color and activating the backlight, or by using an RGB backlight.

A low-cost LCD supplier may have long lead times. If you change manufacturers and order from someone else with the same specifications, one or more segments with the same common may appear dark gray instead of black. To avoid this, specify a lower operating voltage, e.g. 3 V, and use the LCD with 3.3 V. If ghosting occurs, i.e. segments that should not be visible, the voltage can be lowered with a resistor in the supply path or, if the driver allows it, the frequency can be reduced or a longer dead time after the segment update can be selected.

Water jets should not be able to penetrate a unit that meets the International Protection 65 (IP65) standard. If you choose a display that does not come with an IP65-rated mounting sleeve, you will need to add this protection yourself. This is more complicated than it looks. Sealing around the edges of the display (i.e. potting) is too labor intensive. Pressing a sleeve onto the display, which would also mean losing part of the active area, is not possible because the display glass is not designed to withstand such mechanical stress. The only option is to cover the display with a second piece of glass. This leads to the problem of possible condensation between the two panes, unless the second glass is glued to the first, with the glue compensating for the thermal stresses. Bonding is a challenge because you want to avoid air bubbles. If you choose tempered glass for the cover, it is only available in certain sizes at a reasonable price. Tempered glass cannot be cut, so it must be its final size before it is tempered. You may need to have pieces of glass specially tempered. These are all purely mechanical issues. An electrical engineer usually does not have the appropriate knowledge for such a design. It follows that you, as an electronics designer, should not attempt to do this, but should purchase an off-the-shelf IP65-compliant display if needed.

A touch screen on the device and an additional remote control via a smartphone seem to be the state of the art today. However, I found two compelling reasons against these interfaces: humans are haptic, and customers do not want to search for a function in a submenu when using a front panel.

A mechanical switch gives the user immediate feedback when activated. People are used to that kind of haptic feedback. With touch, the feedback must be graphic or auditory. Both are tools. A delay between actuation and confirmation can be very irritating. It is tedious to test that all touch sequences have no latency, if that makes sense at all, because interrupt routines can randomly change latency times.

Touch menus usually result in submenus when displays are not large enough to show all functions at once. A minimum button size is required for large hands. Submenus are easy to create. However, no one wants to wander through the submenus of a washing machine touch display, for example, until they find the function to rinse with more water. They want see all the options immediately, and make a quick selection by pressing a button or turning a dial.

Touch works well on a device that is used almost every day, a smartphone or tablet. You know the response time and you know where everything is by heart. For infrequently used devices, a touch screen often leads to user frustration.

Selecting Off-the-Shelf Modules

Off-the-shelf modules, such as DC/DC converters, motor driver boards, and even entire microcontroller boards, may seem like a convenient alternative to time-consuming and error-prone in-house development. In fact, modules can be the most cost-effective way to achieve this goal in small quantities. However, two conditions must be met.

First, a lifetime purchase must be made (i.e., the purchase of the module in accordance with the total number of units to be produced). The reason for a lifetime purchase could be the risk of discontinuation of the module. This risk is discussed below, but is not essential. The main problem with modules is that they can be changed at any time, unannounced, and above all, without notification from the manufacturer, even if only a capacitor type is changed. However, it is this seemingly trivial change that, when combined with the environment in which the module is used, can lead to a malfunctioning whole. Thousands of units suddenly stop working in the field for reasons that are initially inexplicable and difficult to find. Caution is required with Far East manufacturers. The practice of the "golden sample" is widespread: the manufacturer delivers flawless modules in the first batch, and in subsequent batches, they experimentally test to what extent the savings associated with quality losses are noticed by the buyer. Strictly speaking, a lifetime purchase is not sufficient; you would still need the manufacturer's assurance that all modules come from one batch, and you would have to trust their statement.

Second, the time between the lifetime purchase and the sale of the units produced must be as short as possible, ideally less than a year. Standards are constantly evolving. Modules certified to today's standards may not meet tomorrow's requirements. During a one-year transition period, modules certified to the old standard can still be sold. After that, there are only two painful solutions. Either buy all new modules and repeat all system tests, or have the old modules re-tested to the new standards in the hope that they meet them. The latter can result in significant costs and, more importantly, delays in delivery. It is usually better to buy all new modules.

When modules are purchased and assembled into a system, the following should be noted regarding testing:

- The system integrator must ensure that each module has the required test certifications. If you use a module as intended, you can assume that the tests performed are sufficient.

- However, as a system integrator, you will want to test samples of modules, as well as random samples from production for quality control, according to your own standards. After all, it is your name on the product, not the module manufacturer's.

- The entire system is usually subject to system testing, which is of course the responsibility of the system integrator. If something changes in the module(s), these tests must be repeated.

The best way to integrate modules into your device is to add them separately, connected by wire(s) to your main board. If the module changes, you only need to repeat the system tests. If possible, do not solder them to your printed circuit board. Otherwise, with modules soldered to your PCB, you will have to run all the PCB tests as well if the module changes.

Using Wireless Modules

For wireless modules, the required Specific Absorption Rate (SAR) value for measuring radiation absorption by the user is usually neither measurable nor reasonably estimable by the system engineer. Therefore, this value is usually pre-specified by the module manufacturer. The SAR is calculated for one or more application cases (e.g. for use of the wireless module in a laptop). If the situation does not match the SAR setup, the SAR value is not valid. For example, a laptop radio should not simply be installed in a headset without additional SAR measurements or calculations.

On the other hand, if the module is used as intended and is not connected to a power source, only one test is required for use in the United States. This is the FCC Part 15, Section 15B test. Essentially, it deals with Section 15.109(b) Radiated emission limits of digital devices. After performing these measurements, and if the module is not visible from the outside, the enclosure must be labeled as follows "Contains Radio Module FCC ID: XXX YYYY."

In the European Union, the "Radio Equipment Directive (RED)" simplifies the use of radio modules. The following should be mentioned in this regard:

- RED only affects wireless communications. Modems, telephones and other wired communications are now only covered by the EU Low Voltage and EU EMC Directives.

- Pure receivers must now also be tested for their parameters, including radio receivers, GPS receivers, etc.

- There is no lower test limit for the frequency range; the upper limit is 3 THz.

- Conformity assessment can be performed voluntarily by a notified body on the basis of technical documentation only.

- Customized products for research and development do not require approval, but "evaluation kits" sold to the general public do. In this area, incorrect information is sometimes found on the Internet, probably because the text of the directive is somewhat awkwardly written.

- At trade shows, experimental equipment that is clearly marked as not complying with the Directive may be put into operation in consultation with the national authority.

- Even sub-devices, such as radio modules, which are only functional when installed in the target system, must comply with RED.

- The operating instructions must state the frequency ranges used and the maximum transmitting power.

- Mark of the type, batch and serial number.

The standards under the RED Directive can be found via the EU Low Voltage Directive. If you were to list the standards, you would see that virtually all of them were written by the European Telecommunications Standards Institute (ETSI). Unlike most other standards, the ETSI standards are freely available at www.etsi.org. The standards not only describe the requirements in the band itself (i.e., the maximum power and the modulation method), but also specify the requirements for spurious emissions outside the active radio channel in the entire spectrum. For example, the ETSI standard ETSI EN 300 328 provides the test setup, test configuration and instructions, and the overall limits to be met for the Bluetooth module. Finally, a quick search on the Internet will provide sample test reports.

Do Not Program Apps Yourself

Bluetooth modules encourage the programming of an appropriate application on the cell phone. Although it is fun to program apps, do not do it yourself in a professional environment. You will get into the terrible mess of the consumer race with constantly new smartphones and pads and new operating system versions, where you have to fix all the bugs, even on most of the old versions. Big companies have their own departments for this! I was programming and selling mobile software myself when the first graphic-enabled phones came on the market. Half a year later, I was out of business; there were dozens of different graphic phones on the market, and I would have to port the application to all of them. This constant porting and maintenance of all the old versions needs a group of people just to deal with it. Even a large company should consider whether this makes sense.

There is one exception to this: If you can manage the devices yourself (for example, if you rent out the phones), then the effort required to update an application is clearly limited.

Remote Firmware Update is Complicated

Remote firmware updates for the running operating system of a cell phone are so commonplace today that the uninformed sometimes cannot understand why this should be any different for microcontroller firmware. In my experience, however, a remote firmware update of a running microcontroller is always associated with great effort and contortions, unless it is a Bluetooth microcontroller that has a built-in over-the-air firmware update function. The reasons are as follows:

- Microcontroller memory is usually too small to hold two versions of firmware or libraries.

- A microcontroller without an operating system has no native mechanism to execute functions starting from a different, newly loaded and compiled library.

- Communication modules, e.g. for Bluetooth, do not have the ability to program Flash memory.

- The communication module does not check the integrity of the data and does not calculate a checksum; this is usually the task of the connected microcontroller.

- Communication modules do not have enough memory; the update would have to be imported directly from the received data.

- The usual, quite expensive solution is a second small microcontroller with external Flash memory, both used only to program the main microcontroller from the communication module.

BOM Consolidation

Once the schematic is complete and the components have been determined, go through the BOM with the goal of reducing the number of BOM lines. This is less about reducing component cost and more about reducing production cost and production time. A common example of this is having two different values for pull-up resistors, such as 10 kΩ and 47 kΩ. This can happen if the schematic was created from different reference schematics from different sources. It is often technically irrelevant which of the two values is used. It is not the case that you save on component costs by selecting one of the values for all pull-ups. The difference is that the manufacturer has to use two reels instead of one. If something goes wrong, he has to use the second reel on an additional machine. For no good reason, the production cost is suddenly much higher, which also affects the production time.

In general, it is best to reduce the number of BOM lines as much as possible. If there is free space on a pick and place machine, additional reels can be mounted with the highest number of components per board. This reduces the number of times the machine must be stopped to change reels. The following tips can be used to reduce BOM lines:

- Think about the set of resistors you have. Do you really need all those individual values? Harmonize as much as possible. If you have room on your board, you can even connect resistors in series or in parallel to reduce the number of values.

- If you need a power resistor, you can achieve this by connecting several signal resistors with a positive temperature coefficient in parallel. This will automatically distribute the power among the resistors.

- Reuse a power diode for a signal diode. Reuse a power transistor for logic. Reuse an operational amplifier for a comparator, see the section above. Reuse a linear regulator for a voltage reference.

Visits by Component Vendors

To conclude this chapter, I would like to talk about visits from component salespeople, whether they are a distributor or a manufacturer's representative. You know they're going to be promoting their new products, and you may think it's a waste of time, but consider this:

- You learn about specialty components that you wouldn't have found at the distributor because you weren't looking for them, but that may be the solution. You do not have to choose the latest IC, there will probably be a more established one of the same type available.

58 Component Selection

- In the case of a manufacturer's representative, you will learn about specialty components that distributors do not stock due to low volume.

- They can give you an estimate of how long a component will be available, because they do not want you to be disappointed by a sudden discontinuation.

- Thanks to the personal contact that has been established, you will usually receive samples on request, without having to answer the question of the size of the future order.

- Each representative reveals the problems of the components represented by others.

However, some caution is required:

- Free samples may be "golden samples," i.e., not yet in production and therefore potentially much better than the eventual production parts.

- Representatives must not be given too much insight into the development unless they have signed a non-disclosure agreement, otherwise they may talk to someone else about your projects.

2 Direct-Current Supply Design

With the exception of the next section, this chapter deals with the DC supply. The reason for this is that discussing the design of power (AC) supplies would fill a book of its own. The following section explains why power supply design is so complex and gives advice on how to deal with it.

Power Supply

Designing your own power supply is much more difficult and potentially dangerous than most people realize. It is impossible to even think about a power supply without a grid simulator for brownouts, harmonics, dips, and frequency changes. You need a device that can simulate all the non-idealities of a real public grid, such as Teseq's NSG3040A Multifunction Generator. Typical costs are $600 to rent an older grid simulator for four weeks; the latest model costs about $50 per day, or $6000 for four weeks, with purchase costs in the $16000 range. If your power supply is going to be used worldwide, you need to consider worst-case scenarios. Amplitude and phase can vary greatly. Short interruptions of 500 ms are common. The shape of the mains voltage can be distorted by local sources. Overvoltage peaks and energy-rich, slower overvoltages ("surges") must be able to be intercepted. Designing a power supply is also challenging on the load side:

- The output voltage should turn on and off without overshooting.

- Output current shall be limited.

- Regulated output voltage capable of responding to rapid load changes adequately.

- Low residual ripple for DC output voltages.

- Output voltage should not exceed nominal voltage for light loads.

Not enough, the power supply must fulfill the following requirements:

- The device must not cause excessive radiation interference and must comply with the relevant EMC standards.

- The equipment shall have the lowest possible conversion loss.

- The equipment shall be mounted in a standard enclosure.

- The cooling of the equipment must be properly designed.

Towards the end, when it comes to the seemingly trivial issue of fusing the power supply, you may again be surprised by the considerable effort involved. This is especially true if you are designing a combined 110 V and 230 V power supply. Although UL recognizes fuses as "UL recognized" according to the European standard, this means that additional tests must be performed at UL for certification. Conversely, a UL recognized fuse does not automatically meet the European standard. It starts with the basics: the UL rating of a fuse is the maximum operating current, while the European rating is the melting current. Therefore, you cannot simply buy a fuse that is both UL listed and meets the proper European standards. The environment determines the requirements for the fuse, and these can be so different for UL and EN that it is difficult to find a fuse that meets all the requirements [R007].

Finally, building your own power supply involves high voltages and large amounts of energy. The design is potentially life threatening. For products with small quantities, it is not worthwhile to have your own design; it is much more worthwhile to design so that a standard power supply can be used. For high-volume products, you can get a much better deal than their list price from power supply manufacturers.

Whenever possible, position the power supply as an external power supply outside of your electronics. If the type changes, the whole device does not have to be re-tested, and you can play with different power supply types.

If you still want to stick with your own power supply design, R067, chapter 36 is a good place to start.

There are two main classes of power supplies: linear supplies and switch-mode supplies, shown in Table 16, each with advantages and

	Advantages	**Disadvantages**	**Cost**
Linear power supply	No frequency components > 100 Hz ("quiet").	Heavy transformer Efficiency 50–60 Frequency dependent, input voltage dependent.	Approx. $ 1.50 per watt ranging from 50 to 200 watts.
Switched power supply	Lightweight, small transformer Efficiency 80%-90%.	May cause interference on the power line, on the printed circuit board, and in the environment.	

Table 16. Properties of linear and switched-mode supplies.

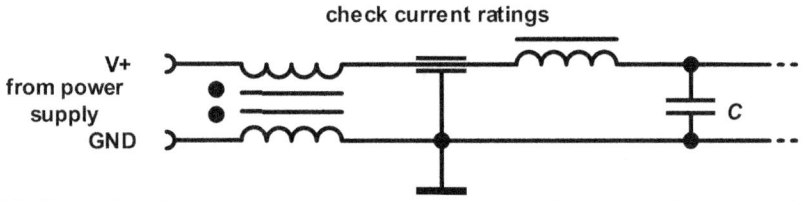

Figure 16. Example of using a common mode choke, a T-filter and a single choke on the supply interface. Shown without ESD protection.

disadvantages. You should not write off linear power supplies per se because of their size and weight. On the contrary, they are preferable for sensitive analog electronics and A/D conversion greater than 12 bits. This is especially true if the power supply is located inside the device, because in the immediate vicinity of a pulsed power supply there may be much higher interference levels than allowed by EMC tests.

After the power supply, the use of a common mode choke is sometimes recommended. An example is shown in Figure 16; see R098 for example values for the circuit components. However, the choke should be designed to be bridgeable, e.g. with 0 Ω resistors. I know of cases where the common mode choke has actually worsened the quality of the supply voltage rather than improving it.

If there are more than a few meters between the power supply and the board, additional protective measures may be required, which are described in the next chapter on Robust Interfaces.

Fuses

What is the worst possible outcome of product development? It is not the non-functioning of the product or its complete failure in the market. The worst possible outcome of product development is human injury. This outcome is not unlikely. In addition to the stove, there are many powerful electrical appliances in a modern household. Today there are even more. Because many portable devices use flammable lithium-ion batteries, almost any electronic device is a potential fire hazard, not just a mains-powered device. The more units sold, the more likely it is that one of them will be involved in a fire. When a fire occurs, every piece of electrical and electronic equipment in the fire area is examined. The first step in this inspection is to determine if the equipment was actually designed and/or approved according to the standards. If the inspection reveals that a piece of equipment has not been tested to all applicable standards, negligence is exposed with costly consequences for the manufacturer, even if the equipment had nothing to do with the cause of the fire. An immediate recall of all equipment of a particular type would follow.

Standards limit design freedom, which is why they sometimes seem like a nuisance to engineers. However, from what has been said above, standards are the engineer's professional life insurance. You do not have to learn to love standards, but every engineer should develop a vested interest in checking his or her design for compliance with applicable standards and insist on it, especially to superiors. Otherwise, you are assuming part of the project risk without financial compensation.

The obvious basic measure against fire hazards is a fuse. However, this solution is not as simple as it seems. If a fuse is required by code, the requirements in Europe, for example, are often different from those in North America. Even the required size may vary. This means two assembly variants, possibly just because of the fuse.

If the standards do not require a fuse, it does not always make sense to include one. An example is a PCB that is powered by a power supply that has overcurrent protection. Typically, the power supply's overcurrent protection is reversible and based on a so-called "hiccup" mode. In this mode, the power supply attempts to increase the output voltage, but immediately shuts down if the short circuit persists. In overcurrent mode, the average current is very low. There is no risk of fire. Adding a fuse does not make the unit safer. A powerful Li-ion battery can be a fire hazard. But again, there is no need for a fuse. You need a charger IC anyway, and these ICs have already implemented much better-defined overcurrent protection than any fuse could provide.

There is one situation where a resettable PTC-based fuse makes sense: if users can attach their own peripherals to your product. Such an attachment may short-circuit the supply voltage you provide. Your board should work again when the faulty attachment is removed.

As an alternative to one-shot fuses and PTCs, you can consider electronic fuses, known as eFuses. These generally have much better defined characteristics. On the other hand, they are a specialty IC, which can cause procurement problems.

The discussion here was conducted from the perspective that the device itself, or a module connected to it, triggers a short circuit. For protection against conducted overvoltages and associated overcurrents, see the next chapter on Robust Interfaces. Fuses are not sufficient protection in these cases.

Reverse Polarity Protection Ideas

Negative voltages can instantly destroy logic ICs. Some measures should be taken to prevent the possibility of negative supply voltages on positive rails. However, Figure 14 shows a reverse polarity protection that you may have learned about in school, but that you should not use. A Schottky diode was used because of the low voltage drop of this type of diode. However, a significant amount of valuable battery power is still wasted once a few milliamps flow. The B140, for example, has a voltage drop of 0.3 V at 10 mA, leaving only 1.2 V in Figure 14. The power dissipation is 3 mW, which seems small. However, 3 mW is 20% of the total power dissipated by the circuit, and that is just to prevent possible reverse polarity!

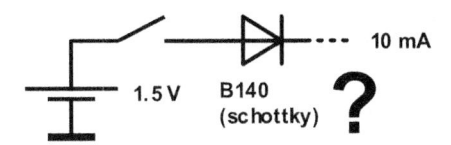

Figure 14. Reverse polarity protection with a PN diode in series.

Figure 15 shows reverse polarity protection with a very low voltage drop. Here an n-type MOSFET (NMOS) is used "upside down," the voltage V_{DS} is negative instead of positive as usual. Note:

- The resistor protects the gate of the MOSFET from ESD.

- The Zener diode is required when the battery voltage is greater than the allowable gate-source voltage V_{GS} which is 8 V for the Si3460DV.

- The 1 MΩ resistor allows the gate to discharge quickly enough when the source voltage can suddenly drop below zero and become negative. Otherwise, the transistor may still be conducting and would allow current to flow in the wrong direction.

- The circuit could also be made using a p-type MOSFET (PMOS), but these have a higher channel resistance. However, in a system where the PCB ground is connected to ground, a PMOS must be used instead of an NMOS. Otherwise, an NMOS could be bypassed by a ground fault.

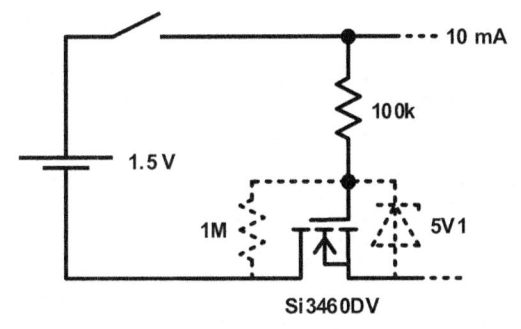

Figure 15. Reverse polarity protection using the Si3460DV NMOS FET. This transistor has an $R_{DS(on)}$ of 41 mΩ at a V_{GS} of 1.5 V and 10 mA; that is, a 0.41 mV drop across the turned-on FET. The optional 1 MΩ resistor and the Zener diode are explained in the text.

- The circuit does not protect the battery from a charge current if one is possible from another source in the circuit, such as a USB connection. The circuit in Figure 15 is therefore not a complete replacement for a diode, because once the transistor conducts, it does so in both directions.

- As the 1.5 V battery approaches its knee voltage of 0.9 V, i.e., its end-of-discharge voltage, $R_{DS(on)}$ rises to 358 mΩ and the voltage drop across the transistor becomes 3.58 mV. This is not much, but it is still more than at the nominal voltage of the battery, so you should do the math with the lowest allowable input voltage.

More advanced reverse-polarity protection solutions, known as ideal diodes, are available. For example, the LM74502H, combined with two back-to-back N-channel MOSFETs in the high-side path, operates from 3.2 V to 65 V. It blocks negative voltages down to -65 V, has an enable pin and 1 μA shutdown current, adjustable over-voltage and under-voltage protection. However, the cost and availability of such specialized ICs must be monitored.

Dealing with Inrush Currents

Batteries and rechargeable batteries (hereafter referred to as "batteries"), even small ones, can often deliver higher short-circuit currents than power supplies. When using battery power, always check the inrush current and make sure it is acceptable. High inrush currents can damage tantalum capacitors and DC/DC converters. This must be determined by measurement. For example, if there is a significant lead length from the battery to the circuit, its inductance will prevent too steep a current rise. These are some other tips:

- Place the capacitors away from the battery connection on the PCB. The large capacitors in the microfarad range will affect the entire board anyway.

- Instead of a tantalum capacitor, use a different type of capacitor that can handle the inrush current (e.g., a ceramic capacitor).

- Do you really need a large capacitor right at the battery terminal?

- Use a choke. This is only possible if the load variations during normal operation are not too high, otherwise large capacitors are required after the choke.

Battery and Accumulator Selection

When using battery power, the question sometimes arises as to whether the battery voltage should be used directly as the supply voltage. After all, any voltage conversion involves additional cost, space, and power dissipation. In addition, some digital components can operate over a wide voltage range. For example, the MSP430F2012 microcontroller operates from 1.8 V to 3.3 V, and a micro-SD card operates from 2.7 V to 3.6 V.

To answer this question, let us first consider the AA alkaline cell, a common battery type. From an open-circuit voltage of about 1.5 V, it exhibits a continuous voltage drop during discharge down to a knee voltage of about 0.9 V, below which the voltage drops more rapidly (see Figure 17). This means that the microcontroller mentioned can be operated almost optimally with two alkaline cells in series due to their voltage range of about 3 V to 1.8 V. The SD card, on the other hand, does not make sense to operate with two alkaline cells.

Now consider a CR2032 coin cell; like other lithium manganese dioxide coin cells, it is a popular source for portable mini-electronics. Figure 17 shows a sample discharge curve. The open circuit voltage of about 3.2 V drops almost immediately to about 2.8 V under load. The knee voltage is about 2.5 V. Thus, the voltage remains virtually constant over the entire useful discharge time. The MSP430F2012 ultra-low-power microcontroller can be operated in the best possible way with one cell; the battery is optimally utilized. Running an SD card directly from a CR2032 button cell is critical. First of all, energy must be stored in a sufficiently large capacitor, because writing to an SD card sometimes requires about 100 mA. But even if this is taken care of, since the knee voltage is well below the SD card's minimum of 2.7 V, operation with one cell would not reasonably drain the battery. Two

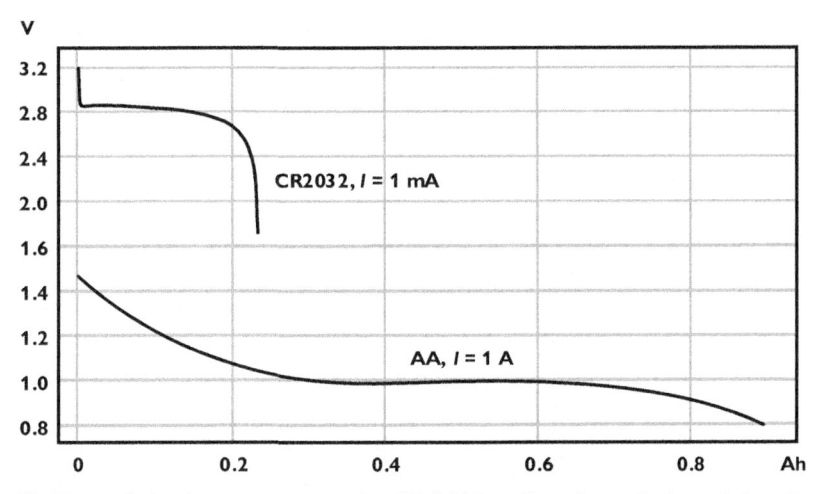

Figure 17. Typical discharge curves of a CR2032 cell and an alkaline AA cell.

CR2032 cells in series are not a direct solution either; together they provide too high a voltage for an SD card.

In addition, the common lithium batteries made of $LiCoO_2$ (3.6 V, classic 18650 type), $LiMnO_2$ (3.7–3.8 V), $LiFePO_4$ (3.3 V) or Li_2FePO_4F (3.6 V) or then as lithium polymer batteries (almost all 3.7 V or 4.2 V) deliver a voltage too high for direct connection of most ICs or too low for 5 V technology. It is of little use that they all have the same basic flat discharge curve as the CR2032 battery shown in Figure 17.

NiMH batteries decay under load from 1.5 V to a flat discharge curve at about 1.2 V for a long time before reaching the knee voltage at 1.1 V. The MSP430F2012 microcontroller works best with two NiMH cells. However, with three NiMH cells, the SD card gets too little voltage at the beginning.

In conclusion, it is almost a stroke of luck to be able to power your electronics directly from a battery or rechargeable battery and use the available charge completely. With rechargeable batteries, it may be tolerable to stop discharging before the knee point, which means a shorter runtime and more charge cycles overall. However, runtime is usually an important consideration, so even with rechargeable batteries, the goal is to get as much energy out of the battery as possible.

Before we turn to the methods for accomplishing what we just said, a word about alkaline and NiMH cells. Let us suppose that some device users use NiMH batteries instead of alkaline batteries without being aware of the lower voltage. When the device stops working, they are surprised and disappointed. Therefore, for maximum customer satisfaction, when designing battery-powered devices, consider using existing rechargeable batteries of the same size as a working alternative.

Batteries and Voltage Regulators

In general, it is not possible to connect electronics directly to a cell or cells and extract optimal energy. The next simplest solution is a regulator. However, you are burning a lot of energy in the regulator itself, which is absurd given the precious charge of the source. But there are cases where you may not be able to use the much more efficient DC/DC converters shown below, e.g. because these switching regulators introduce noise into the circuit and there are problems with sensitive analog devices such as receivers or certain sensors.

For example, if you are forced to use a regulator, you would need four alkaline cells with a combined knee voltage of 3.6 V, which is just enough headroom for a low-dropout regulator. Therefore, the idea of using a 9 V block battery is not far off, especially since its knee voltage is between 6 V and 7 V. This is enough to fully utilize the battery beyond the knee point with a control margin of about 1.7 V of a conventional L78L33 regulator. Initially, however, with a fresh battery, a 5.7 V drop across the regulator indicates an efficiency of only 36%. At knee voltage, the efficiency is no higher than 50%. Meanwhile, a 9 V battery has a low capacity of about 500 mAh for its size. Three AAA alkaline batteries in parallel are only slightly wider than the 9 V block, but flatter (i.e. they have about the same volume), and have about five times the capacity of the 9 V block. Together, these batteries are about half the price if the user does not buy the 9 V battery in a multipack.

The 9 V battery is also tricky to install. It was the solution to power the popular transistor radios of the 60s. These 9 V batteries are still used in smoke detectors today because the ionization chambers available are designed to operate on 6 V and the piezo speaker needs to sound loud enough. In addition, a smoke detector is a pure analog circuit with a current draw of about 40 µA, so it will work for a year on one battery. For conventional digital circuits, however, the use of a 9 V battery is unreasonable.

With low dropout regulators (LDOs), there is a risk of misapplying components, as shown in Figure 18. The LP2985AIM-3.3 regulator datasheet has specific requirements for input and output capacitors (see Table 17). This includes an equivalent series resistance (ESR) specification for the output capacitor to prevent the regulator from oscillating. Classic linear regulators are designed to never oscillate, but they require a significant margin in the difference between input and output voltages. The problem with the example in Figure 18 is that the

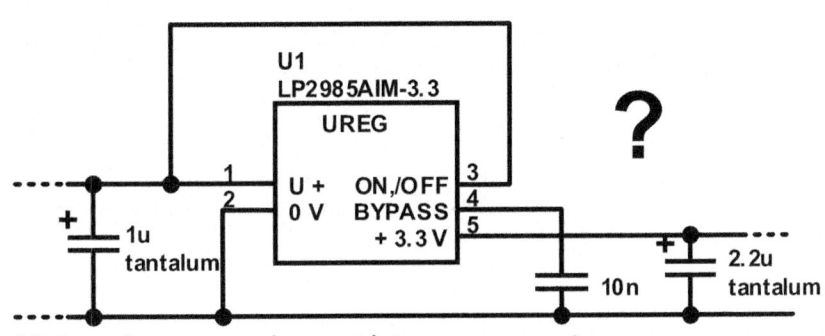

Figure 18. Low dropout regulator with incorrect capacitors.

Variable	Input capacitor	Output capacitor
Minimum capacitance value over all parameter values especially temperature	1 µF	2.2 µF
ESR range	Not relevant	5 mΩ–800 mΩ

Table 17. Input and output capacitor requirements for the LP2985 controller.

input and output capacitors were chosen incorrectly. One possible reason for using tantalum capacitors is that the capacitors shown in Figure 18 have been used elsewhere as power supply decoupling capacitors on printed circuit boards. Therefore, they are likely to be reused. Another possible reason for the use of tantalum capacitors could be that the first generation of low-dropout regulators were designed specifically for tantalum capacitors, to the point that a saying has become common among engineers: "Low-dropout regulators hate ceramic capacitors". Discussion of the schematic in Figure 18 reveals the following insights:

- A low dropout regulator must always be specified as a "ceramic type" or "tantalum type" and the capacitors selected accordingly.

- For critical minimum capacitance values, add a sufficient margin to the theoretical value; see capacitor selection below.

- The output capacitor should be placed as close to the controller as possible. I know of one case where a regulator with the correct output capacitor started to oscillate because it was 1 cm away!

LDOs are typically available as bipolar or complementary metal-oxide-semiconductor (CMOS) types. Bipolar LDOs have 10 times less noise than CMOS types and are more efficient at higher voltages and currents. CMOS LDOs often have digital features such as sleep mode or low battery alarm output.

Note that the dropout voltage at higher currents can be quite high, even with an LDO. While it is only 110 mV at 100 mA for the LM2941, using the same LDO at 1 A results in a voltage drop of 0.5 V. However, this is still a small margin compared to about 2 V for a non-LDO typical linear regulator.

Batteries and Charge Pumps

Only DC/DC converters can achieve the standby and operating times per battery charge expected of today's portable devices. The simplest DC/DC converter ICs are charge pump ICs, e.g. the TPS 60100 for a regulated 3.3 V output voltage: the input voltage can vary from 3.6 V down to 1.8 V, i.e. each of two alkaline cells in series can be used down to 0.9 V, which corresponds to their knee voltage. The average efficiency is approximately 70%. The IC has a quiescent current of 50 µA and a standby current of 50 nA. The conversion frequency is fixed at 300 kHz. Of course, this conversion comes at a cost. The IC itself costs about $1 each for 1000 units, plus two 2.2 µF, one 10 µF, and one 22 µF capacitor, and together they take up board space. Without regulation, these costs will be passed on to the customer, who will have to recharge the device more often or replace the batteries more often.

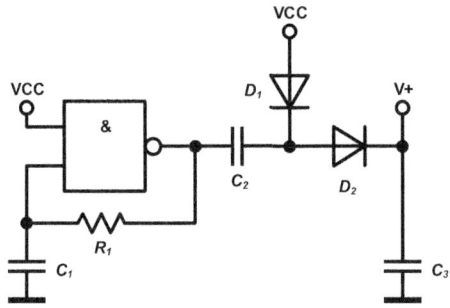

Figure 19. Example of a voltage doubler circuit for a few milliamps.

The cost of a DC/DC converter IC may tempt you to design your own simple discrete charge pump. The circuit shown in Figure 19 is well known for a few milliamperes. Despite the apparent simplicity of the circuit, it is strongly discouraged; the cost/benefit ratio is simply too low. Where an IC is available, its price/performance ratio is usually better. Otherwise, no one would buy such ICs and everyone would build their own discrete solutions.

Boost Converter for Single Cell Operation

Charge pump ICs are only available up to about 100 mA. For higher currents, inductive converters are used. In this case, step-up or boost converters allow operation from a single cell. An example is the LM2623 with a possible input voltage of 0.8 V to 14 V and a higher output voltage, fixed at 1.24 V to 14 V, with a maximum current of 2 A. The quiescent current is 80 µA and only 2.5 µA in standby mode. The IC sells for about 50 cents per 1000 pieces and requires a 4.7 µH coil, three capacitors of up to 100 µF, a diode, and three resistors. The switching frequency is selectable from 300 kHz to 2 MHz, which is useful if you have a problem with a particular frequency or its harmonics.

Batteries and a Step-Down Converter

If you need more power, you connect several cells in series and usually get above the operating voltage, at least with fully charged batteries or accumulators. If the combined knee voltage is as high as the system voltage, a buck converter can be used optimally. For example, this would be the case with three alkaline cells and an operating voltage of 2.7 V. However, if the knee voltage is lower than the operating voltage, the result is sometimes unfavorable. With three alkaline cells and a 3.3 V system, a pure buck converter would indicate the end of operation at this voltage, but the combined series knee voltage only occurs at 2.7 V. If the battery voltage were linear with decreasing capacity, you would use (4.5 V–3.3 V)/(4.5 V–2.7 V) × 100% = 67% of the battery energy before indicating that the batteries are "discharged." However, in Figure 17 a few pages back, you can see that the alkaline cell voltage drops faster at first, before changing to a quasi-linear curve. You can see from the graph that at 1.1 V (i.e., one-third of the three alkaline cells in series for a system voltage of 3.3 V), only about 50% of the energy has been extracted from the batteries. Of course, the customer pays the price and usually does not notice that the batteries are not fully discharged. However, half the operating time compared to a competitor's product that uses a buck-boost converter and all the available battery energy can be noticed by any layman. If the device is rarely used and 50% of the usable energy is consumed by standby current, the user will have to insert new batteries every time he or she wants to use the device. Finally, it makes no ecological sense for users to throw away half-full batteries. In my opinion, if a battery or accumulator is not fully utilized, it is a hidden defect. Such failures have the potential to rebound on the person responsible for them at the most inopportune moment.

Batteries and Buck-Boost Converters

What you can lose by using a buck-only converter attached to a battery was shown in the previous paragraph. Here we discuss the use of a buck-boost converter and its advantages. An example is the BD8303MUV from ROHM, with a variable input voltage from 2.7 V to 14 V, a fixable output voltage from 1.8 V to 12 V, and a maximum current of 1.5 A. The quiescent current is high at 1 mA, but the device draws a reasonable 1 µA in standby. The switching frequency is selectable from 480 kHz to 720 kHz, so if you have problems with switching noise at a particular frequency or its harmonics, you can change it. The IC costs about $1.50 each at 1000, and it also requires a 4.7 µH inductor, four FETs, two diodes, seven capacitors of up to 47 µF, and four resistors. This means a higher cost and a significant increase in space compared to the buck-only converter in the previous paragraph. But, as mentioned above, you would waste nearly 50% of the battery power with the pure buck converter. Instead, with the buck-boost converter, the full battery charge is usable.

DC/DC Converter Selection

Choosing a DC/DC converter IC can seem like a Herculean task: so many parameters to consider, pages of datasheets to study. Fortunately, this is no longer the case. With webench.ti.com from Texas Instruments, webdesigner+ from OnSemi, and similar online platforms from other manufacturers, you can enter your requirements, and solutions are suggested, including the complete circuit and even the layout. There is no reason not to take advantage of this, unless you are tied to a DC/DC IC manufacturer that does not offer this service.

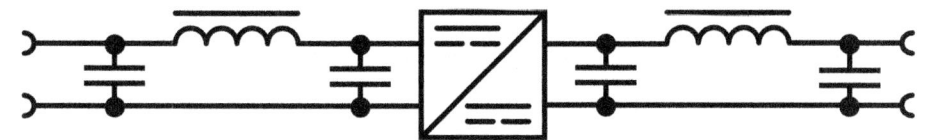

Figure 20. DC/DC converter with input and output pi filters.

DC/DC Converter Input/Output Filtering

Often pi filters are added to the DC/DC converter, either before or at the output, or both, see Figure 20. When is this useful? Are there any detrimental effects of adding a filter? To answer these questions, let us consider the example of a typical buck converter, as shown in Figure 21, and analyze the currents and their waveforms:

- At the input, there is a pulsed current with high amplitude.

- At the output there is a millivolt voltage ripple.

These may lead to the following issues:

- Does not pass EMC test because too much is coupled out or radiated away. This is especially true if the supply line to the converter is long. To mitigate this, we use an input filter.

- The switching frequency and harmonics ghosting around the supply are rectified by op-amp input protection diodes and can cause distortion in sensitive analog circuits. We use an output filter to mitigate this.

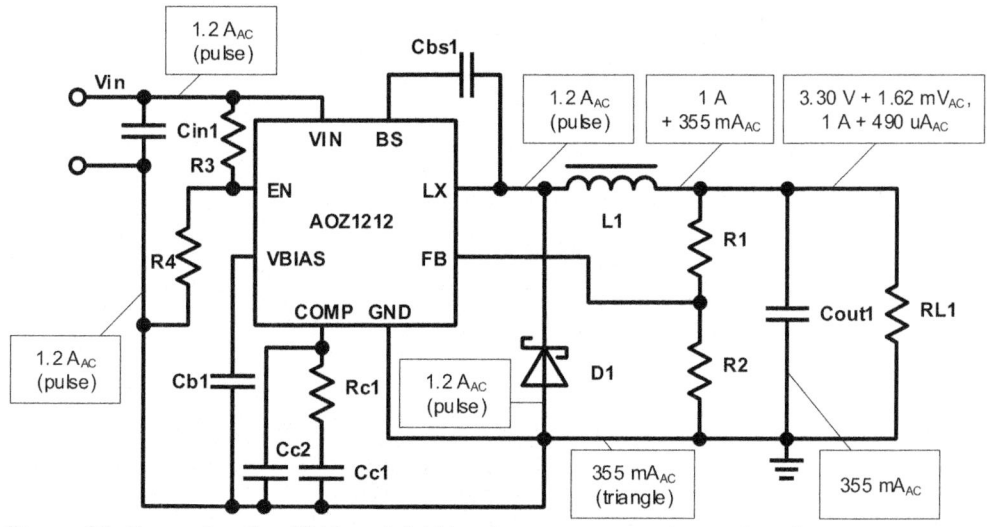

Figure 21. Example of a 12 V to 3.3 V buck converter and associated currents.

72 Direct-Current Supply Design

Do not use an output filter if the load is digital, especially if you are driving high speed devices. An output filter degrades the ability to deliver high currents quickly. The result is increased noise on the digital supply.

Do not use pi filters if your equipment is subjected to vibration and shock, as the ferrite core of the coil is brittle and prone to breakage.

Pi Filters: Notes on Chokes and Ferrite Beads

A pi filter contains an inductive component. This is usually a choke or a ferrite bead, the latter being nothing more than a choke with a few turns. To understand how a choke works in two ways against harmonics, see a typical characteristic in Figure 22:

- In the example, the choke acts almost as a pure inductance up to 10 MHz. Core losses have virtually no effect on the total impedance, which increases by 20 dB/decade. At 4 MHz, an inductance of 100 Ω can be read, resulting in an inductance of 4 µH, which can be used in conjunction with the capacitors to form an LC filter.

- Above about 10 MHz, losses in the core become dominant. This results in a further increase in impedance. That is, the filter continues to operate as if inductance were present, with the only difference being a different phase response.

- Above 1 GHz, the ferrite choke begins to behave capacitively, and the impedance begins to decrease by 20 dB/decade. A choke is therefore a broadband device.

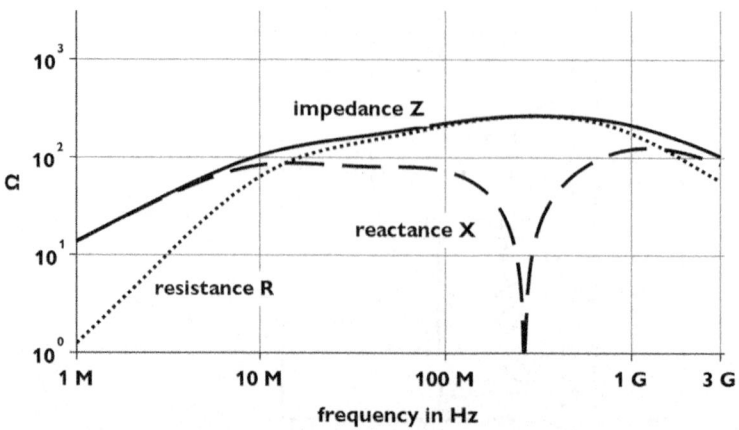

Figure 22. Frequency response of the ferrite choke BLM18RK221SN1.

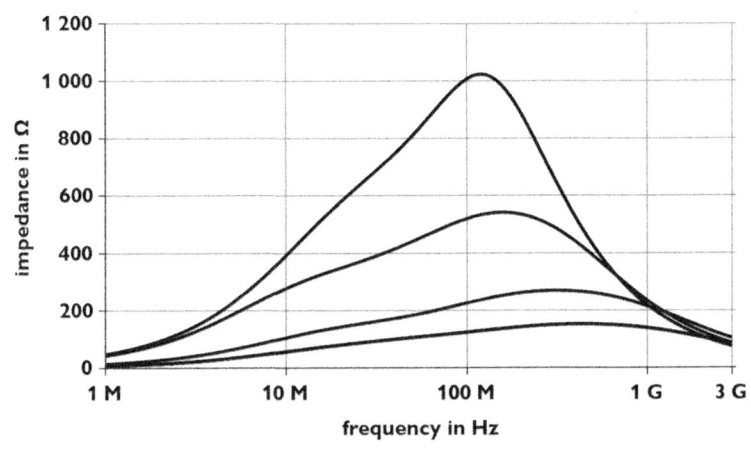

Figure 23. Frequency response of BLM18RK chokes with 120 Ω, 220 Ω, 470 Ω and 1 kΩ @ 100 MHz.

Chokes are typically specified with the maximum of the impedance (Ω @ Hz). Due to the high losses of the choke at the resonant frequency, no ringing occurs there. Comparatively, the losses are also high at lower frequencies, so the resonant circuit consisting of the choke and a nearby decoupling capacitor or input capacitance is irrelevant. This is true as long as the current does not saturate the choke. To safely avoid this, select an inductor with a maximum current approximately twice the supply current. Figure 23 shows the impedance curves of various Murata BLM18RK series chokes. The operating frequency range of approximately 10 MHz to 1 GHz is the same for all. But how do you select the right type for a particular application? It turns out that the best way to select a choke is to have several types at hand and to evaluate them on a trial-and-error basis together with the capacitors, as explained in the next paragraph.

As a hint, choose chokes in standard packages such as 0805 that can be elegantly bridged with a 0 ohm resistor when not needed or even counterproductive.

Pi Filters: Notes on Capacitors

If you decide to implement a filter, start with a choke type and select the capacitors to obtain a filter cut-off frequency that is 10 times lower than the switching frequency. Common switching frequencies of DC/DC converters are 150 kHz or 300 kHz, so we get cut-off frequencies of 15 kHz or 30 kHz, respectively. At such frequencies, chokes act as nearly ideal lossless coils, and you can use their inductance value to find the proper capacitor values. Note that the input capacitor(s) and the output capacitor(s) become part of the filter. Choosing a cut-off frequency that low gives us 40 dB of attenuation at the switching frequency and 40 dB/decade higher.

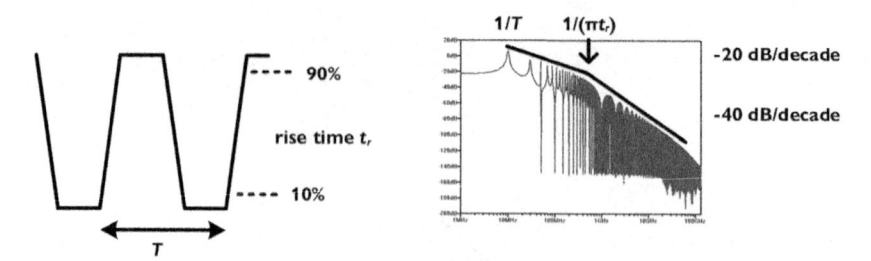

Figure 24. Spectrum of a trapezoidal signal with 50% duty cycle. For better representation, the rise time is set to 17% of the clock frequency.

Remember from the chapter on component selection that our capacitors have a limited bandwidth within which they will operate as desired. So the question immediately arises: up to what frequency does a DC/DC filter need to operate?

First, it helps to determine the spectrum of a trapezoidal signal, see Figure 24. This is a good model for describing pulsed DC/DC current waveforms. The spectrum of a trapezoidal signal has two prominent points:

- The first peak in the diagram appears at the switching frequency.

- The transition of the envelope from -20 dB/decade to -40 dB/decade happens at the so-called knee frequency equal to $1/(\pi \times \text{rise time}) = 0.32/\text{rise time}$ [R095].

How much filtering is necessary depends on the specific situation, of course, but in general it is necessary to filter at least up to the knee frequency. After that, the harmonics themselves decrease by 40 dB/decade anyway.

Often the switcher rise time is unknown. The following rule of thumb can help: if you do not know the rise time, it is best to assume that the rise time is 7% of the cycle time [R089]. This results in a knee frequency that is approximately five times the switching frequency. This means that the capacitors must be capacitive up to five times the switching frequency.

An LC filter can have resonance overshoot. However, since the resonant frequency of the filter is selected to be 10 times lower than the switching frequency, this is not a problem at first: there is no excitation from the DC/DC converter. However, if there is a strong signal in the system at exactly this frequency, it can cause the filter to oscillate. For this reason, it is always necessary to have the option of bypassing the coil or replacing it with a 0 Ω resistor.

Applying the above rule of thumb to a 300 kHz DC/DC converter results in a knee frequency of 1.5 MHz. With microfarad capacitors, this becomes tight; their resonant frequency is just in the range of 1 MHz and below. Meanwhile, the choke losses are not yet dominant. We still

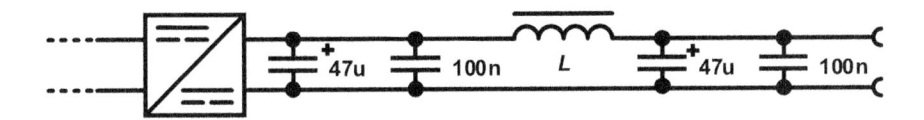

Figure 25. Extending the bandwidth of a pi filter by adding MLCCs.

have to rely on the attenuation of the LC filter. What we can do is to add some nanofarad ceramic capacitors in parallel with the microfarad electrolytic ones, see Figure 25. If we do this, it is important that the choke does not get too much current, otherwise the high-Q ceramic capacitors may oscillate with the choke, which becomes a high-Q air coil when saturated. Consider the following additional notes on capacitor selection:

- The maximum value of an output capacitor is determined by the maximum capacitive load of the DC/DC converter.

- Capacitance values that are too high will cause startup problems due to excessive inrush current. Occasionally, this may result in attempted startup, especially if the capacitor is connected to the output without a choke.

- Note that all other decoupling capacitors must be added to the output capacitor, especially the microfarad capacitors.

Pi Filter Variants

Another suggestion is the filter shown in Figure 26. It illustrates the method of providing several assembly variants without unduly bloating the layout, which must be kept compact. For example, a 1206 ferrite choke can be replaced with a 0 Ω resistor if the filter is not used or if the choke's influence is negative, as described above. Also shown is the use of a ferrite choke in the ground path. This prevents common mode currents in the ground.

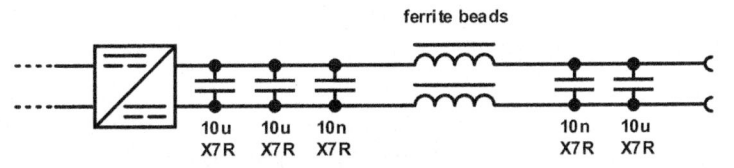

Figure 26. Pi filter with a ferrite in the supply path and one in the ground path.

76   Direct-Current Supply Design

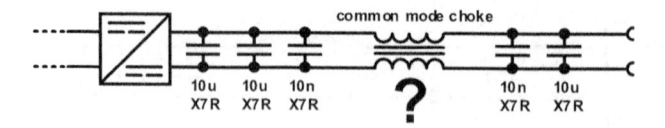

Figure 27. Non-functioning pi filter with common mode choke.

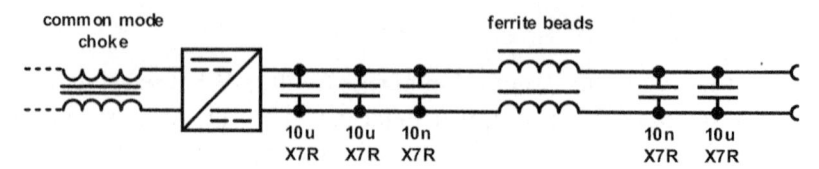

Figure 28. Common mode choke at the input of a DC/DC converter.

When discussing the example in Figure 26, it is important to note that using a common mode choke instead of two single ferrite chokes, as incorrectly done in Figure 27, will destroy the filter. A common mode choke has a vanishingly small inductance for a differential mode signal. However, the noise of the DC/DC converter has a differential mode character.

Practical experience has shown that a common mode choke at the input of the DC/DC converter (Figure 28) has solved many EMC problems. Therefore, I recommend placing a common mode choke on each DC/DC converter input.

DC/DC Converter Layout

The layout of DC/DC converter circuits is often a source of uncertainty. Unfortunately, the converter IC data sheets are usually of limited help: they show an example, as in Figure 29, but it is not a fully implemented layout. Trying to follow the example exactly often fails due to the size of the components, see Figure 30. In a way, this is intentional. If there were a complete layout example in the da-ta sheet, it would probably be picked up in many cases, but then changed just as quickly without understanding what the critical points are.

These are the critical DC/DC converter layout points to manage:

- Identify the traces that carry high switching current. In the case of the buck converter shown in Figure 21 a few pages back, this is the Vin-IC-diode-GND path back to where Vin enters the board. Keep this trace as short as possible, on one side of the board, and with a minimum loop area. Figure 30 is a good example of the implementation of this principle. From the connector, the Vin voltage goes directly to the switcher IC. Right at its switching output is the free-wheeling diode, which is already pointing back to the supply input. There follows a straight line to the connector.

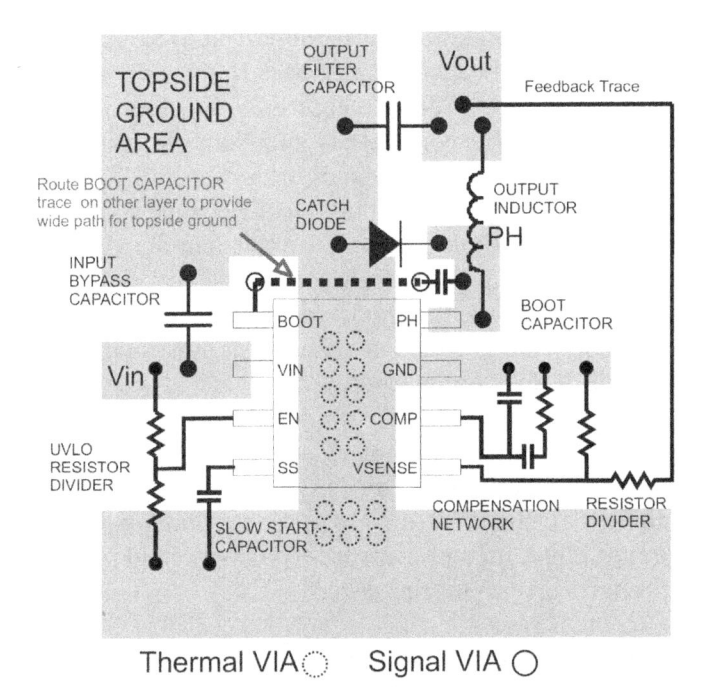

Professional Electronic Design Best Practices 77

Figure 29. Layout recommendations from the TPS54331 IC datasheet.

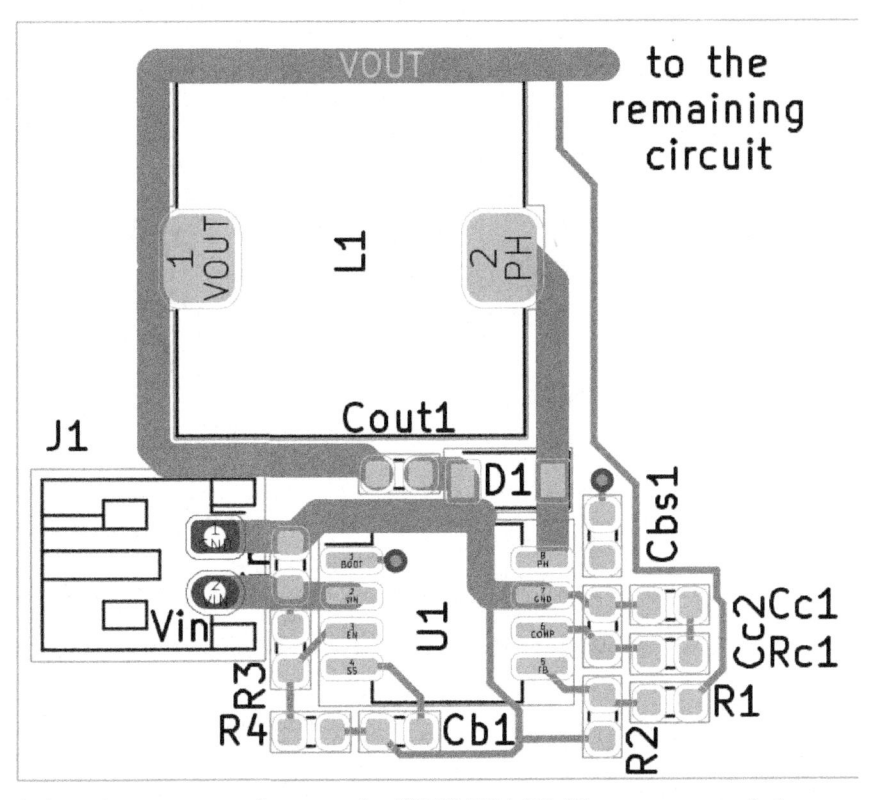

Figure 30. Real layout example using the TPS54331 IC. The recommendations given in the datasheet could not be fully followed to save space, e.g. the coil had to be placed on the top left instead of the top right.

- The second, but much less critical, loop to control is the one from the coil to the output capacitor and back to the coil, whether via a diode or a switched MOSFET. In Figure 29, the datasheet suggestion, and Figure 30, the complete layout example, this path is kept as short as possible, avoiding traces under the coil. Tracks under large components are always a bit tricky because you cannot see if their solder mask is damaged, and the space under the components tends to pick up potentially conductive dirt. After all, it makes no sense to have the tracks to and from the coil underneath. The coil itself usually has many windings, so if you add one more outside, it does not matter at all.

- As explained above, I would not place the feedback wire under the coil either, but it can be placed right next to it without problems, especially with a shielded coil. Magnetic shielding works best close to the shield; further away, the magnetic field reaches the same strength as without the shield.

- Keep everything else close together, all tracks as short as possible, avoid using another layer except for low current tracks.

If you have difficulty designing a layout that adheres well to the above points, it may simply be due to an inappropriate pin order of the DC/DC converter IC, see Figure 31 for an example.

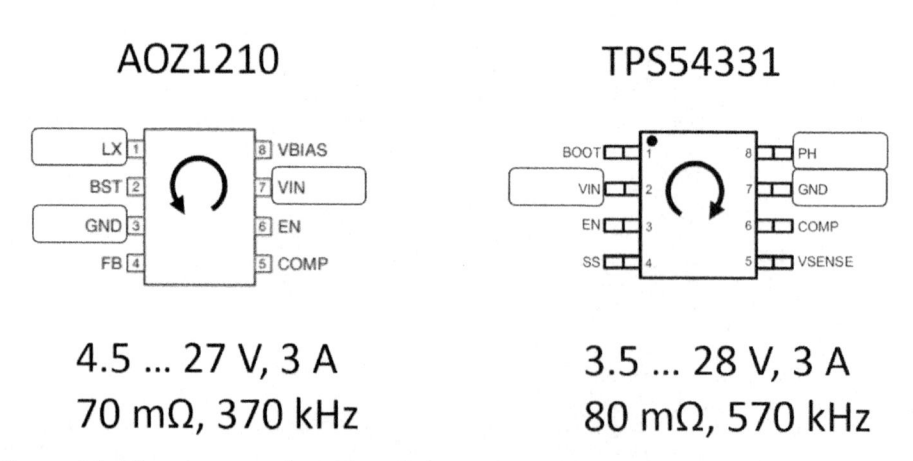

AOZ1210

Pins: LX (1), BST (2), GND (3), FB (4), COMP (5), EN (6), VIN (7), VBIAS (8)

4.5 ... 27 V, 3 A
70 mΩ, 370 kHz

TPS54331

Pins: BOOT (1), VIN (2), EN (3), SS (4), VSENSE (5), COMP (6), GND (7), PH (8)

3.5 ... 28 V, 3 A
80 mΩ, 570 kHz

Figure 31. The two switcher ICs AOZ1210 and TPS54331 have similar basic characteristics, but their pin-out order regarding Vin, LX/PH and GND are opposite. Depending on the environment, one may result in a more complicated layout than the other.

Low-Noise Switchers

The alternative to a conventional DC/DC converter and output filter is a low noise switching regulator. The LT1777 data sheet provides a standard circuit that is guaranteed to produce a very low noise DC voltage. However, this limits component selection and substitution.

Cascaded, Series, and Parallel DC/DC Converters

Cascading DC/DC converters is generally unproblematic, but the following must be observed. If the inrush current of the second DC/DC converter is too high, the first DC/DC converter may not start. A control circuit, as shown in Figure 32, allows the second DC/DC converter to start with a delay if the first DC/DC converter does not have a Power Good output [R066]. If the load is not applied to the second DC/DC converter right from the start, a capacitor C_1 in Figure 32 is large enough to provide the inrush current when the load comes on. In any case, it is a good idea to buffer the intermediate voltage strongly so that you are more immune to load fluctuations.

Be aware, that not only complex ICs like certain FPGAs but simpler ones too might require an exact startup sequence of the supply voltages. In one project, the DAC IC AD7305 broke regularly. When we studied the data sheet more thoroughly, we stumbled upon the sentence "It is recommended to power VDD/VSS first before applying any voltage to the reference terminals to avoid potential latch up". We found that sometimes the reference voltage was actually there first, before the supply voltage. There was no means implemented to control the voltage sequencing.

If multiple voltage starts must follow a sequence, the solution shown in Figure 32 using simple RC delay circuits can still be used. However, the seemingly simple alternative of using Power Good–Enable connections must be thoroughly analyzed for validity when more than two voltages are involved. Power Good outputs are typically open drain. If we were to route the Power Good signal from the OKL to the EN5322 switcher in Figure 33, both ICs would still try to start simultaneously. The OKL IC needs some voltage and time to turn on its power

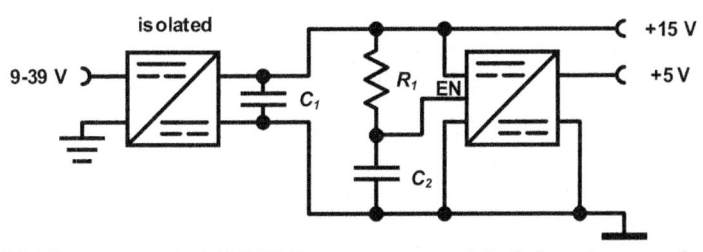

Figure 32. Two cascaded DC/DC converters with delayed start of the second converter if the first converter does not have a Power Good output [R066].

80 Direct-Current Supply Design

Figure 33. Example of the use of a start-up sequence IC instead of Power Good to Enable connections, because the latter would lead to a race condition in this case, see text for explanation.

good transistor to signal that the output is not yet stable. For complex cases, it is better to use a dedicated start-up sequence IC, as shown in Figure 33. The start-up sequence IC also has the advantage of a correct power-down sequence. However, it is advisable to have an alternative type that can be mounted. Nothing worse than not being able to produce the board due to a missing startup IC.

In series connection, the output of one DC/DC converter is connected in series with the output of another DC/DC converter (see Figure 34). It is essential that the individual outputs are filtered. Otherwise, low beat frequencies may occur that are difficult to filter out. If a converter fails (e.g. due to thermal overload and automatic shutdown), it becomes part of the load, and a negative voltage is applied to its inputs. This negative voltage causes a current through the body diodes

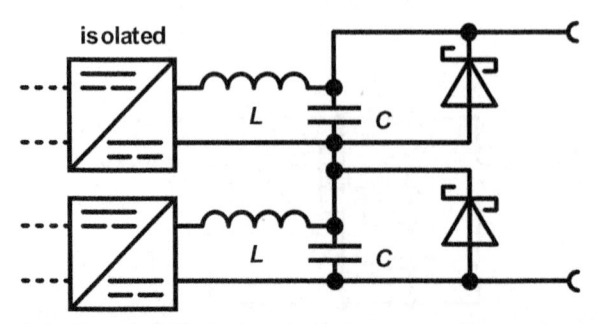

Figure 34. Two DC/DC converters in series. Schottky diodes are used for protection in the case of failure of one converter (see text).

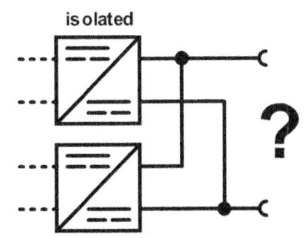

Figure 35. Two DC/DC converters in parallel.

in the output stage. This can destroy the output stage of the inactive converter. The diode must be able to carry the load current for as long as it takes for the system to detect that a converter has failed.

Parallel connection of DC/DC converters, as shown in Figure 35, is not possible because there is no "load balancing" (i.e., load sharing). Overloading of one of the two DC/DC converters cannot be excluded. For a parallel connection, the DC/DC converters must have a special load sharing function, which requires an additional connection between the two converters. Alternatively, load sharing controllers must be used to distribute the load equally between the two converters.

DC/DC Converter Limitations

A high inrush current can damage the DC/DC converter and cause EMC problems. There are ways to prevent this:

- An inductor of 1 mH or greater on the input. Note: Filter chokes are typically in the range of 50 to 300 µH and may not be sufficient to limit inrush current.

- When a common mode choke is used, the common mode component of the inrush current is suppressed, which is beneficial.

Provide each DC/DC converter with a filter at the input. Measure the inrush current, e.g., via the voltage across the coil $I = 1/L \times \int V\,dt$.

When using a DC/DC converter, it is usually necessary to choose between the two main types of common output current limiting (see also Figure 36):

- Constant Current. The maximum output current is kept constant by decreasing the output voltage, even as the load resistance shrinks. In the extreme case, the load becomes a short circuit and the total power is equal to the output voltage times the maximum output current developed in the controller. The controller must be able to handle this, and the heat sink must be designed for this. Expect a large heat sink.

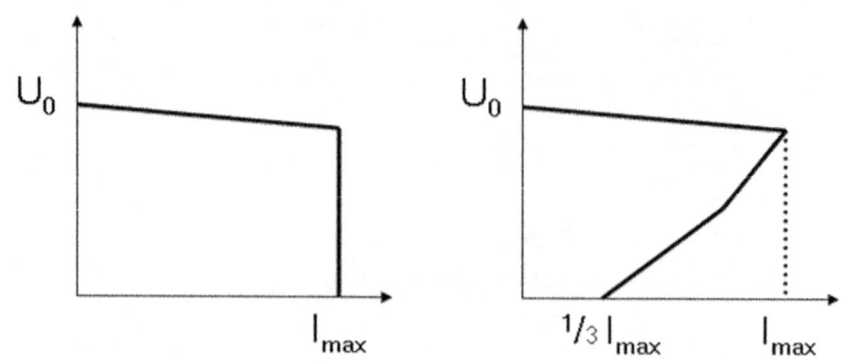

Figure 36. Constant current limiting on left, foldback protection on right.

- Foldback Protection. The output current is reduced according to the foldback characteristic. After the maximum output current is reached, the load resistance decreases. In the event of a short circuit, the current flowing at a normal load is significantly less than the maximum output current of the controller. In terms of power dissipation, the heat sink only needs to be designed for the maximum output current flowing at the nominal output voltage. There is a drawback to foldback protection: if the total capacitance of the decoupling capacitors is large and the output voltage at the controller rises too quickly on power-up, excessive current is drawn temporarily and the controller enters foldback mode, delaying start-up.

Some converters have short-circuit protection in the sense that in the event of a very high overcurrent, the converter will completely shut down the output, or will only allow a minimum current to pass.

Clocked regulators sometimes have a "hiccup" mode for this purpose: a pulse is sent to the output from time to time to determine if the short is still present.

Some converters have protection against excessive junction temperature by shutting down the converter at a certain measured junction temperature. This overtemperature protection usually has no hysteresis. If the heat sink is too small, the converter may turn on and off in a slow cycle of a few seconds.

Current Measurement Variants

Currents often need to be monitored in electronic devices. Direct measurement via a low-impedance shunt allows the use of multiple pin- and function-compatible ICs. In addition, the variants shown in Table 18 are available [R149]. A few more comments on current measurement:

- Measurements with a shunt and instrumentation amplifier are susceptible to noise. Using a microcontroller with a fast ADC and fast processing, the analog signal can be read in with an excessive clock rate and then averaged. This is especially recommended for motor current measurements because of the strong noise field [R149].

- In the case of a low-side measurement, the negative potential of the driven element is not equal to the ground potential. For example, the potential of the metal motor housing is different from that of the ground plane. Such differences cause stray currents and EMC problems.

- If the current to be measured changes very quickly, e.g. 1 A/ns, check the inductance of the shunt. If the inductance of the shunt is too high, the current calculated from the resistance of the shunt will be incorrect.

- Measurement error must be calculated at minimum current; see R149 for details.

- Currents in low-power devices often cannot be measured using a shunt soldered to the PCB because the operating mode currents are in the milliampere range, while the standby mode currents are in the microampere or nanoampere range. To get volt readings in both situations, the shunt must be changed. This can only be done with an external measuring device connected to a current track that is split, e.g. by a jumper.

Of course, any measurement has to be enabled first, as shown in Figure 37. This often means having separate power nets, even though they carry the same voltage: for example, if you want to check that the microcontroller is really in sleep mode and that it only draws the current associated with it, you need to separate the microcontroller's power supply tracks from the tracks leading to other ICs.

IC at shunt	Comment and example
Single supply, rail-to-rail op-amp	Can be used for low-side measurement only and is inexpensive. The parasitic resistance between the shunt and ground falsifies the result; it can vary considerably in production e.g. due to different copper purity in the via wall copper plating.
Differential amplifier	Input voltage >> IC supply voltage is possible. Wider input voltage ranges than CSMs (see below). Obtains the essential current from the measured source (i.e. only applicable for large currents). Continuously adjustable gain, example INA 146.
Instrument amplifier with voltage output	Typical application: Input voltage ≤ IC supply voltage. Output voltage range may be severely limited (INA826), but not necessarily (INA326). Gain factors can only be selected from one set.
Instrument amplifier with current output.	The gain factor is variable. High capacitive load is possible by connecting an appropriately compensated op-amp. Internal trimming to the required absolute value means an expensive IC.
Current shunt monitor.	Input voltage >> IC supply voltage possible. Fixed gain; example: INA 282.
Digital current measurement ICs.	Resolution is predefined, shunt must be selected accordingly.

Table 18. Several options for measuring current with a shunt.

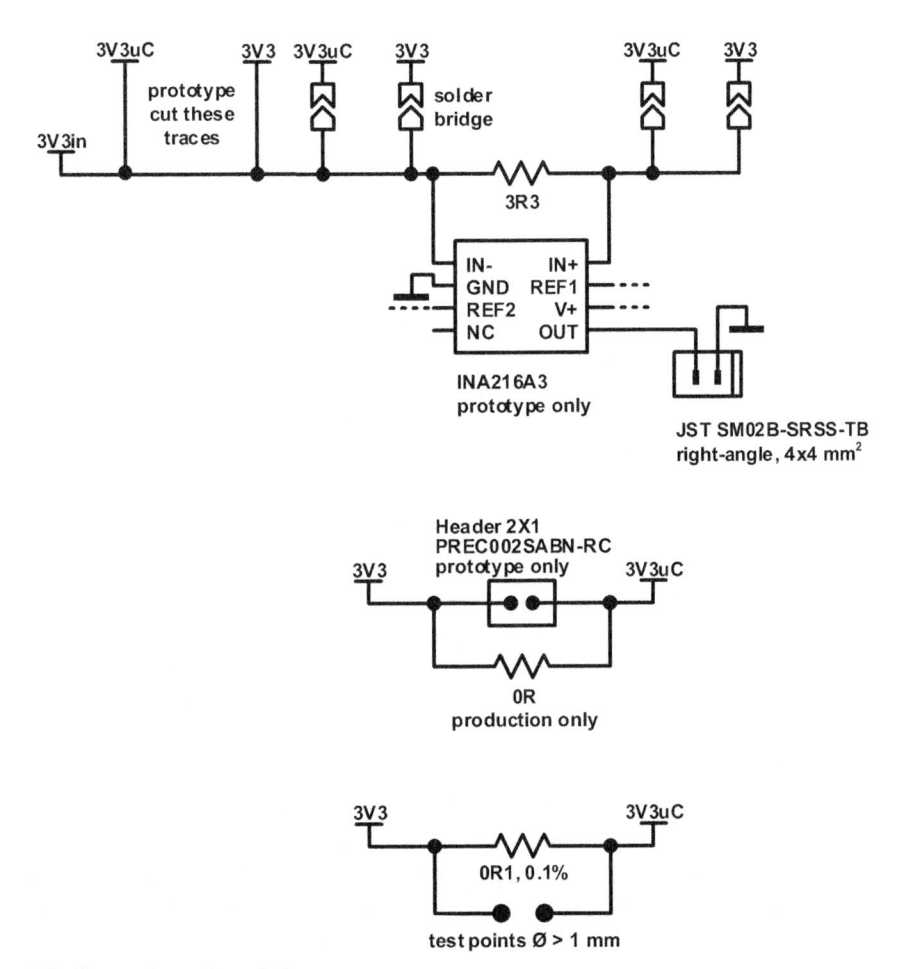

Figure 37. Examples of enabling current measurements.

Maximum Currents in Traces and Vias

The main concern with excessive trace currents is overheating, which can lead to lift-off or burn-through of the PCB tracks. The IPC-2221 design guideline provides some guidance on this [R014]. However, the curves in the IPC-2221 document are copied from the older IPC-D-275 or MIL-STD-275 sources [R013]. According to the curves, 400 mA at a trace width of 6 mil = 0.15 mm and a 35 μm layer thickness results in a 20°C temperature rise on an external trace. Note that if you choose 18-μm copper foil for your board, you will end up with approximately 35 μm of copper height on the outer layers. The process of plating via walls also deposits copper on the outer layers—typically around 17 μm. Check with your board manufacturer for exact values. The IPC-2221 charts do not take into account the plating process, they are based on the final copper height.

Tin plating barely increases the thermal cross section of a track, as its thermal conductivity is 10 times worse than that of copper. Although the thermal conductivity of gold is only slightly worse than that of copper, gold plating does not significantly increase the thermal cross section due to the thin layer.

MIL-STD-275 allows a maximum temperature rise of 20°C, but temperature rises of 30°C to 40°C are common [R012]. A value to remember is half an ampere per 0.15 mm trace at 35 μm copper thickness. Traces on inner layers have the same temperature rise, according to IPC-2221, at 1/2.6 = 38.4% of the trace width on the outer layer. However, a study has shown that internal layers can now carry considerably more current due to denser circuitry, and a different treatment than for outer layer traces is not justified [R004].

The use of a power plane eliminates the issue of continuous current capability. A power plane with a 35 μm copper layer has a typical sheet resistance of 0.5 mΩ/square. This results in a negligible heat dissipation of 50 mW/square, even at high currents such as 10 A. The same is true for currents in the ground plane.

However, the fact that both the supply and return currents often pass through vias raises the question of the continuous current carrying capacity of the supply and ground connections. The resistance of an unfilled via is highly dependent on the wall thickness. This wall thickness can vary considerably from manufacturer to manufacturer. The copper used is also less pure than on the layers [R068]. With the available data, the resistance can be calculated [R069], but there are also studies on the subject that give a value of about 2.5 mΩ and a maximum load of 1 A to 3 A [R070].

Maximum DC Source Impedance

With digital logic in CMOS, the noise of the power supply and the ground appear one-to-one on the digital signal. A noisy supply or a noisy ground therefore means a reduction of the noise margin up to an unstable system. Therefore, clean power and ground are very important in digital systems. Let us take the example of the Cortex-M4 microcontroller STM32F429. Although it allows an operating voltage of 1.7–3.6 V, there must never be a difference of more than 50 mV

Symbol	Ratings	Min	Max	Unit
$|\Delta V_{DDx}|$	Variations between different V_{DD} power pins	-	50	mV

Figure 38. Excerpt from the STM32F429 microcontroller datasheet regarding the allowed voltage difference between two supply pins. With a maximum possible current draw of 100 mA through a supply pin, this results in a maximum source impedance of 0.5 Ω.

between any two of the 16 supply pins, as stated in the "Absolute Maximum Rating" part of the datasheet (see Figure 38). At the same time, each of these supply pins can draw up to 100 mA of current and the device can draw a total of 270 mA across all supply pins.

We do not know the spectral content of the supply noise. Therefore, the worst-case supply impedance at a given supply pin must not exceed 50 mV/0.1 A = 500 mΩ at any relevant frequency.

For the relevant bandwidth where the maximum supply impedance should not be exceeded, the formula f_{max} = 0.32 / rise time can be used according to Figure 24 earlier in this book. If the rise time is unknown, 7% of the period is a good estimate [R089]. For the STM32F429 operating at 180 MHz, this results in an estimated rise time of 389 ps and a bandwidth of 918 MHz. In this example, the supply must stay below 0.5 Ω up to nearly 1 GHz! How can we achieve this?

Power Supply Decoupling Without a Power Plane

Before we begin this discussion, a quick note on terminology:

- We talk about "decoupling" when we want to create a low-impedance supply that can deliver the necessary current to the ICs without a voltage dip.

- We talk about "bypassing" when we need to route more or less undefined noise to ground, a typical situation is smoothing the voltage after a DC/DC converter.

- In practice, these two terms are often used synonymously, but to be precise, they are not.

Suppose it is not possible to dedicate a PCB layer to a power plane. Often thin tracks are used for power distribution when currents are small. However, a track 0.2 mm wide and 35 μm thick has a resistance of about 25 mΩ/cm. A trace 20 cm long, e.g. from an IC to the nearest microfarad capacitor, would already reach the 0.5 Ω limit of the supply impedance mentioned in the previous paragraph. If such long traces are required, the width of the traces should be increased significantly to reduce their resistance. On the other hand, if all the power traces are short, as in Table 19, you can still use thin traces to save space.

The next concern may be via resistance. There is considerable variability due to different wall thicknesses, so the best we can say is that vias have resistances on the order of 2.5 mΩ. This is small enough to be ignored in most cases.

Staying with the vias, let's look at their capacitance and inductance. There is a capacitance between the via wall and the layers, but these two electrodes are perpendicular to each other. The via capacitance turns out to be in the sub-picofarad range and can be neglected. The

88 Direct-Current Supply Design

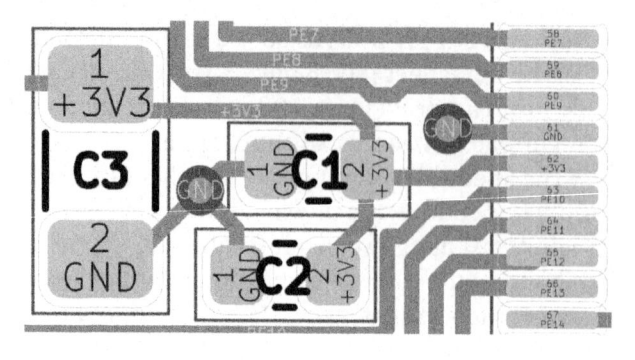

X7R 0603	C	L_{para}	Part and size	C	L_{para}
C1	1 nF	0.5 nH	C3 Tantalum, C	330 µF	3 nH
C2	100 nF	0.6 nH	Via Ø 0.8 mm	≈ 0	0.14 nH (R197)

Table 19. Example of a power supply decoupling without a power plane using decoupling capacitors on two supply pins of the STM43F429. A 1 nF, a 100 nF, and a 330 µF capacitor were placed as close as possible to the supply pin and connected to the ground plane by vias. The ground plane is located on the other side of a standard PCB; an inductance of 8.6 nF/cm was assumed for all 0.2 mm wide traces to calculate their inductance values. The L_{para} values are the parasitic inductances of the parts as explained in the text.

via parasitic inductance is due to the cylindrical current flow, which is about 0.1 nH/mm board thickness for a 0.8 mm diameter via, according to R197.

Going back to the trace length discussion, Table 20 lists the resulting inductance per centimeter for two trace widths and certain heights of those traces above the ground plane. Multiplying one of the values in Table 20 by the length of a trace gives the inductance of the loop formed by the current in that trace and the returning current in the ground plane. Strictly speaking, the values given are not the inductance per centimeter of the trace alone, but of the entire circuit, based on the length of the trace and depending on the height of the trace above the ground plane. Nevertheless, we are talking about the "inductance of a trace". To be more precise, the inductance of the vias needed to connect the trace to the ground plane and close the loop is not included in the values listed in Table 20 and must be added.

Compared to the much larger capacitance of the capacitor, the capacitance of the traces can be neglected.

Calculation method		FEMM	R197	FEMM	R197
(stack-up diagram: 0.36 mm, 0.71 mm, 0.36 mm)	top copper height 35 µm prepreg 7628, 0.18 mm prepreg 7628, 0.18 mm copper 35 µm	nH/cm For w = 0.2 mm 5.3	 4.8	nH/cm For w = 0.4 mm 4.1	 3.4
	FR4 core	-	-	-	-
	copper 35 µm prepreg 7628, 0.18 mm	7.7	7.0	6.5	5.6
	prepreg 7628, 0.18 mm bottom copper 35 µm	8.6	7.6	7.3	6.2

Table 20. Inductance of a trace over a ground plane, without power plane. The resulting values vary depending on the layer of the ground plane. FEMM values obtained using the Finite Element Method Magnetics software [R193].

When it comes to the parasitic inductance of the capacitor, things get a bit tricky. The parasitic inductance L_{para} of a capacitor is measured by manufacturers as follows: a loop is built, and its inductance is measured. Then, a portion of the loop of the exact length of the capacitor is removed and replaced with the capacitor. The loop inductance is measured again. The difference between these inductances is the parasitic inductance L_{para} of the capacitor (R201). This means that L_{para} is the excess inductance in addition to the inductance of a DC path between the capacitor ends. If we now want to find the total inductance of capacitor C1 in Table 19, for example, we take its L_{para} = 0.5 nH and the inductance of a virtual path between the capacitor ends. For the 0603 case size in the example, the length is 1.55 mm and the internal electrode width is approximately 0.8 mm. A trace of this length and width on the top of the board, with the ground plane on the bottom, has an inductance of about 3 nH. This results in a total C1 inductance of 0.5 nH + 3 nH = 3.5 nH.

Next, we add the track inductance from the IC power pin to C1 and the track and via inductance from C1 to ground. Finally, we add the via and track inductance from the ground plane to the IC ground pin.

To summarize the previous sections, Figure 39 shows the resulting effective equivalent circuit for a capacitor connected to adjacent VDD and VSS pins of an IC and placed close to it.

If we repeat this procedure for all the capacitors, each time combining all the inductances into a loop inductance and connecting the three loops in parallel, we end up with the equivalent circuit shown in Figure 40.

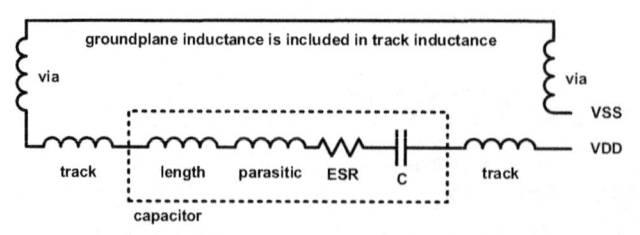

Figure 39. Effective equivalent circuit for the impedance of a capacitor placed as close as possible to an IC and connected to its VDD and VSS pins. Due to the short traces, their resistance can be neglected. The same is true for the via resistances. The small track and via capacitances can also be ignored. See text for an explanation of track inductance, length inductance, and parasitic inductance.

If we calculate the impedance of the decoupling network circuit of Figure 40, we start from 100 kHz with an impedance of about 0.5 Ω, then go much lower—towards 10 MHz—but then shoot up into an enormous anti-resonance, see Figure 41. The anti-resonance results from the resonant circuit of the parasitic inductance of the 100 nF capacitor and the 1 nF capacitor. With a second 100 nF capacitor instead of the 1 nF capacitor, the antiresonance disappears, but the broadband impedance target is still missed. The example shows that the common use of capacitors at the IC supply pins does not provide sufficient decoupling of the supply in the high-frequency range. This is also confirmed experimentally in R121.

Before turning to improved power supply decoupling using a power plane, let us briefly discuss the parallel connection of decoupling capacitors. First, Figure 41 shows that decoupling capacitors are useful beyond their resonant frequencies. This is because the phase of the noise in the supply is not important, only the magnitude. Therefore, it is not necessary to add another decoupling capacitor with a higher resonant frequency in parallel in order to decouple the power supply more broadly. On the contrary, as we have just seen, at certain fre-

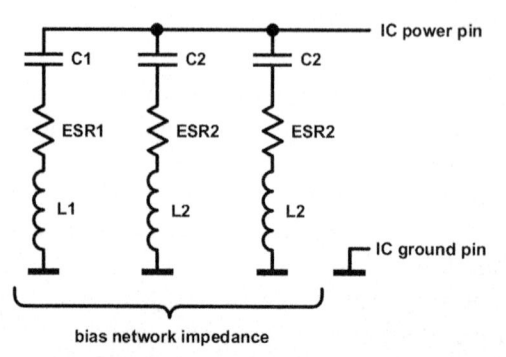

Figure 40. Equivalent circuit for the decoupling network.

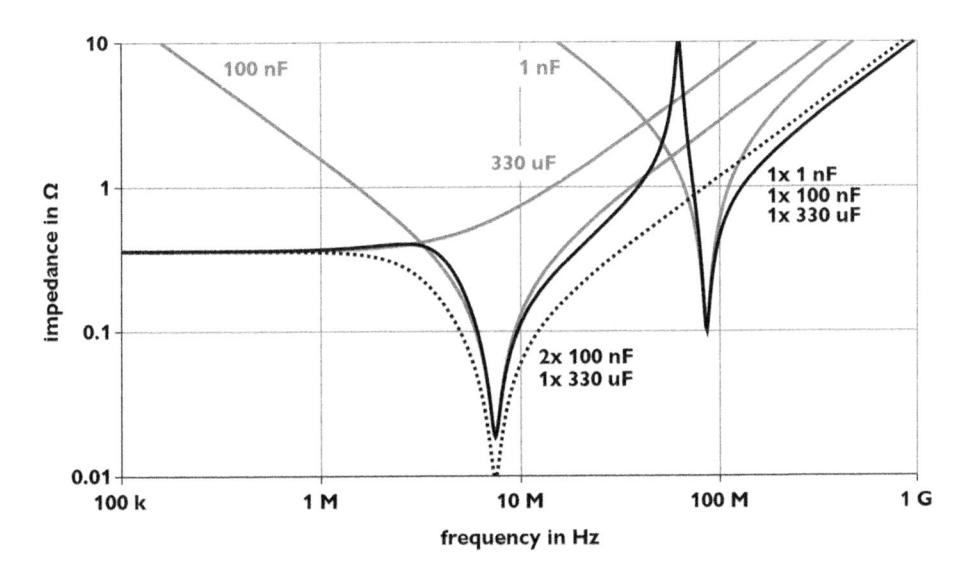

Figure 41. Example of power supply decoupling without a power plane. The black line drawn represents the impedance of a 330 μF tantalum capacitor in a C case in parallel with a 100 nF X7R capacitor and a 1 nF X7R capacitor, each in a 0603 case, including track inductance. The resulting impedance curve is dotted with a second 100 nF capacitor instead of the 1 nF capacitor. The gray lines are the frequency responses of each capacitor.

quencies the supply impedance can be severely degraded by a pronounced anti-resonance. However, this only applies to the case without a power plane. With a power plane the situation is different, as explained below. Hence the idea to do it here without a power plane. As shown in Figure 41, without a power plane it is better to connect capacitors of the same size in parallel instead of a much smaller one; in the example a second 100 nF capacitor in a 0603 case. These parallel capacitors have a combined capacitance of 200 nF, but half the inductance of a single 100 nF capacitor. The parallel connection cannot be replaced by a single 220 nF capacitor in the 0603 case because it would have approximately the same inductance as the 100 nF capacitor. Only by using a single capacitor in a 0201 case would the inductance be significantly lower. However, such small capacitors with comparatively high capacitance can only be achieved with a ceramic material containing a lot of barium titanate, e.g. the X5R type, which leads to numerous disadvantages listed in the chapter on component selection.

The benefit of paralleling capacitors in terms of reducing inductance while increasing capacitance is particularly noticeable when using a power plane, which will now be examined.

Power Supply Decoupling with a Power Plane

Using a power plane instead of a network of traces eliminates the significance of ohmic resistance. For example, an 18 μm thick inner power plane has a resistance of approximately 1 mΩ/square. This applies to a current flowing uniformly through the square, entering at one edge and exiting at the opposite edge. If we feed the square with a current from one point on one edge and drain it from one point on the opposite edge, e.g. with two vias, the resistance increases by a factor of about five to 5 mΩ/square, which is still low enough to be completely negligible. This result holds up to high frequencies, although the skin effect increases the resistance. At 1 GHz, the skin depth of copper is 2.1 μm. If we take this value twice, once for the current flowing at the top of the plane and once for the current flowing at the bottom of the plane, we still have a value of only 25 mΩ/square.

As a rule of thumb, a power plane-ground plane pair separated by a perpendicular distance d has an area inductance of 1.2 nH × d / square. Using the Finite Element Method Magnetics (FEMM) software, a value of 1.3 nH × d / square has been obtained for various values of d [R193]. However, these results are based on a uniform current distribution from a selected edge to the opposite edge. In reality, the current enters the power plane at specific points: at the output of a DC/DC converter, at the positive terminals of the capacitors. A simulation shows that with a point source at one edge of the power plane and a point drain at the opposite edge, the inductances only increase by a factor of about 1.5 to 1.86 nH × d.

If we arbitrarily choose d to be small, we can reduce the resulting inductance, which will be beneficial, as can be seen immediately. As an example, a 4-layer pooled PCB typically has a 0.71 mm FR4 core and 0.36 mm prepreg layers on both sides. If we use the third layer as the power plane and the fourth layer as the ground plane, we get an inductance of only 0.43 nH using a point source and point drain on opposite edges of a square PCB. In reality, the source and drain points are well inside the board area, and in addition to a virtual square shape between the points, there is even more copper between them, further reducing the inductance. The result is such low inductance values that we can consider all supply decoupling capacitors to be directly connected to each other, directly connected to the source, and directly connected to all IC supply pins. This finding is supported by experiments in R121, where the author concludes that all decoupling capacitors can be placed anywhere on the PCB without changing the supply impedance! Note that this is only true if the power plane is close to the ground plane.

Figure 42 shows an example of a direct parallel connection of many decoupling capacitors. This example can be considered again in the application for the STM32F429. There is no 100 nF capacitor at each supply pin, there are only six of them, but there are 12 pieces of 10 nF capacitors, 3 pieces of 1 μF ceramic capacitors, and 3 pieces of 330 μF

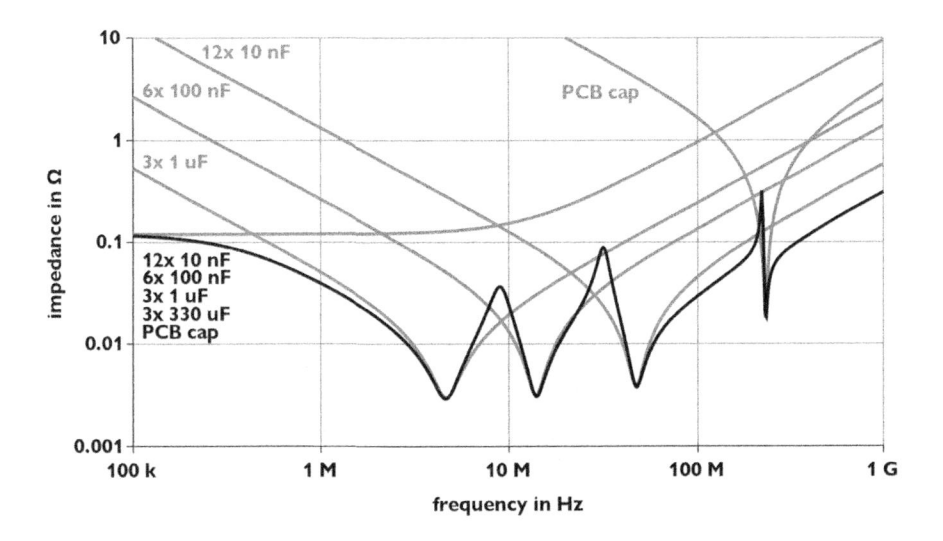

Figure 42. Example of a power supply decoupling scheme with a power plane. The line drawn represents the impedance of three 330 µF tantalum capacitors connected in parallel with three 1 µF, six 100 nF and twelve 10 nF X7R ceramic capacitors, all in 0603 packages, and additionally in parallel with the capacitor formed by the power plane and the ground plane, these being spaced 0.36 mm apart and each having an area of 80 x 100 mm^2. All capacitors were provided with two via inductances of 0.14 nH and a connection inductance of 0.54 nH. A broadband low supply impedance with controlled anti-resonances results.

tantalum capacitors. A via inductance of 0.14 nH to ground was added to each decoupling capacitor, and the same inductance for a via to the power plane.

Since all of the board's decoupling capacitors are now working together, we get a broadband, very low supply impedance of 0.1 Ω up to about 300 MHz. This is below the 50 mV/270 mA = 185 mΩ requirement, where we must use the device's maximum current draw. Figure 42 also shows that the anti-resonances, which are the peaks of the black combined impedance curve, are also below 0.1 Ω. This is by design; the ceramic capacitors follow each other closely in value over a decade and have been selected in number to provide approximately the same ESR value per capacitor group. In order to achieve a certain ESR value, twice as many 10-fold smaller capacitors in the same package were connected in parallel. If the resulting ESR had been too low, the anti-resonance peaks would have been higher. Therefore, decoupling capacitors must be carefully selected in terms of number and ESR. This number-ESR combination can be determined by simulating or calculating the parallel connection of all decoupling capacitors and should be done at least once for a specific circuit.

Since the anti-resonances of the X7R capacitors can be well controlled, the choice of Y5V types is not advantageous. On the contrary, a

94 Direct-Current Supply Design

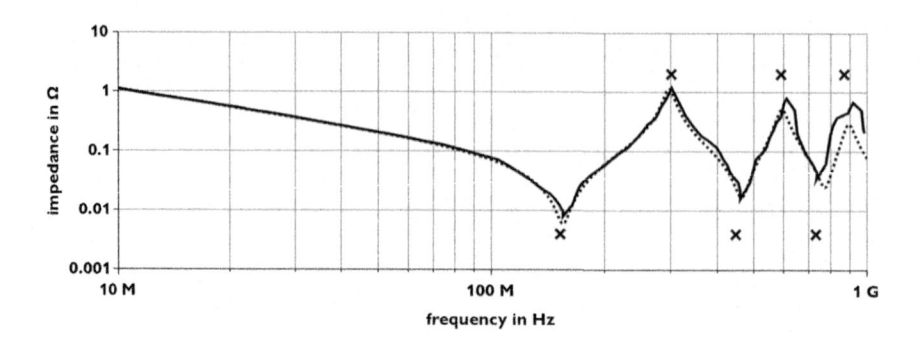

Figure 43. Example of the impedance curve of a power plane-ground plane pair. Both planes are 262 mm × 102 mm, spacing 0.61 mm, relative permittivity 4. Extracted line measurement according to R121, dotted line calculation according to R195, and crossed calculated resonant frequencies according to text with arbitrary assigned impedance.

decoupling system designed for broadband applications, such as the one shown in Figure 42, is at risk due to the high manufacturing tolerances and strong aging of the Y5V types.

From the above illustrations, it is also clear that the rule of thumb "at least one 100 nF capacitor for each supply pin of a digital IC" is quite useful even without a power plane, but its full effect unfolds only with one.

So far we have not discussed the capacitance of the power-ground plane pair. In the example shown in Figure 42, this capacitance is quite small and does not contribute significantly to lowering the supply impedance. On the contrary, it is detrimental in terms of forming an anti-resonance with the inductances of the capacitors. However, the peak of anti-resonance is very narrow band and in practice is much lower due to damping by the lossy PCB materials. We accept this in favor of the low impedance connection provided by the power plane for all decoupling capacitors and all VCC pins of the ICs.

In addition to an anti-resonance peak, there may be another problem when the power and ground planes are large. For example, for a 262 mm × 102 mm board, Figure 43 shows the impedance curve of the pair up to 1 GHz. The peaks in the diagram are not caused by anti-resonance, but are actually caused by the pair itself. What we see is the power plane acting as a patch antenna. The frequencies marked with crosses in Figure 43 are calculated using the Formula $f_{res,n} = n \times L \times c_0 / \sqrt{\varepsilon_r} / 4$, where L is the longest side of the board, c_0 is the speed of light in vacuum, and ε_r is the relative dielectric constant of the FR4 material. They correspond to the self-resonant frequencies of an end-fed antenna of length L. Odd-numbered n describe high-impedance resonances, and even-numbered n describe low-impedance resonances.

In particular, the peaks of high impedance resonances are problematic if there is a signal at exactly those frequencies, e.g. a clock har-

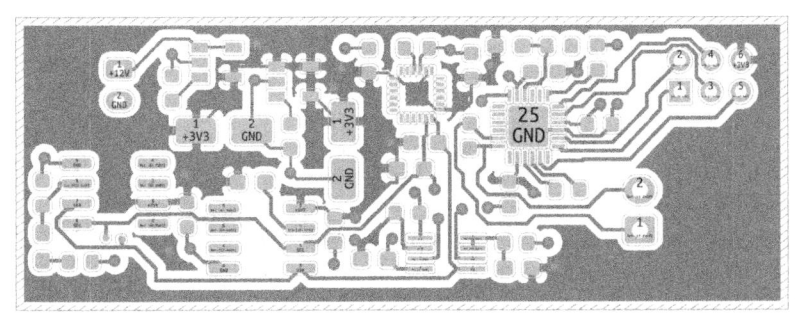

Professional Electronic Design Best Practices 95

|←→| 2.5 mm

Figure 44. Example of flooding the top layer with copper at 3.3 V potential.

monic. Therefore, the self-resonance frequencies of the power-ground plane pair must be calculated from the outset based on the planned board size and set in relation to the clock and signal frequencies, including their harmonics.

The patch antenna resonant frequencies of a power plane do not change with layer spacing, but the directivity does. Only a change in board size will shift the resonant frequencies. A circular power plane acts as a patch antenna with circular polarization. Only an irregular PCB contour will prevent pronounced resonances.

The best solution would be to route fast clocks and other digital signals differentially. Since the charge circulates only between the driver and receiver in a different direction, there is no need to pull charge from the power plane or push it back. There are no back-and-forth currents on either the power or ground plane, and you may be able to save quite a bit on backup capacitors, since a very low supply impedance is not required when using differential signals.

A flooded power plane, as shown for an example in Figure 44, can exhibit propagation time resonances in the same way as a dedicated power plane layer. However, additional filled layers do not add new resonant frequencies, at least not in the range up to 1 GHz, but they do add capacitance. In an example reported in R121, the ground plane power plane capacitance of a 6-layer PCMCIA board was increased from 0.5 pF to 4.1 nF by proper layer sequencing and consistent filling of the potential layers. Only then did the board pass the emission test. So should all layers be flooded as far as possible, as shown in Figure 44? From the point of view of power supply isolation, the answer is clearly "yes". In addition, there are other benefits. Heat dissipation is improved due to the reduced thermal resistance across the layer. The risk of PCB bowing due to uneven copper distribution is reduced.

However, there is a risk of flooding. Crosstalk can be increased. As explained in the chapter on signals on PCBs, the distance between the signals and the flooded areas must be sufficiently large. There must be no potential-free copper islands. Unfortunately, this often means that flooding cannot be done sensibly because it results in an overly fragmented and unconnected patchwork.

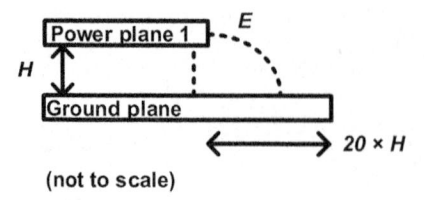

Figure 44. 20H rule regarding board edge and power plane extension.

When flooding areas, pay attention to the thermal connection of the elements to the plane. For example, if both pads of a capacitor are heated differently, one side can lift up during soldering, the so-called tombstone effect.

Finally, it is recommended that you follow the 20H rule with regard to board edges. The 20H rule states that there should be a distance of 20 × H between the end edge of the power plane and the nearby edge of the ground plane, where H is the height of the power plane above the ground plane, see Figure 44. This is based on the fact that approximately 70% of the stray field is then contained within the PCB area. This is considered a reasonable compromise. At 100 × H it would be 98%, but also a correspondingly large reduction in power plane size.

Decoupling Capacitors and the Time Domain

If you have a bad feeling about placing the decoupling capacitors anywhere on the PCB when using a power plane, as explained in the paragraphs above, you are right. An impedance calculation is only valid for a steady state. Statistically, such a state exists when the noise in the supply is purely random and relatively small. This is usually the case when many gates switch virtually randomly. However, it can happen that the number of gates switching on is much larger than the number of gates switching off at a given clock pulse. With such a statistical "outlier," the above derivation is no longer valid.

Let us look again at the example with the STM32F429 microcontroller and a sudden maximum possible current draw of 100 mA at a certain VCC pin with an estimated rise time of 389 ps. With a typical conduction time of 15 cm/ns, the current rise information during the edge itself has a horizon of 0.389 ns × 15 cm/ns = 5.8 cm. Capacitors further away cannot contribute to the decoupling of the supply.

Hence the conclusion: Yes, with respect to the impedance of the power supply, the placement of the decoupling capacitors hardly matters if a power plane is present. However, with respect to irregular current draw, it is still best to place the decoupling capacitors as close as possible to the IC supply pins [R105]. The next question is where to place the via to the power plane. As shown in Figure 45, it is often possible to place the via on the other side of the pin under the IC—the optimal solution. If you are not allowed to place tracks under the IC, you will usually not be able to place the first decoupling capacitor as close

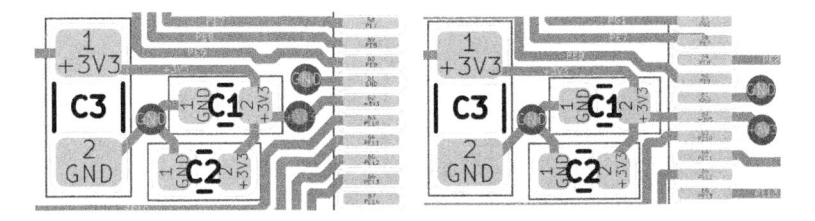

Figure 45. Power plane via and decoupling capacitor placement options.

to the pin as you would like because the adjacent pins will not be accessible. This leaves room for a via to the power plane. The solution with the via directly at the IC's supply pin is also preferred by some engineers because the via-power-plane combination has a much lower inductance than the decoupling capacitor. For very fast load changes, it is then easier for the IC to draw charge, but only a limited amount is available, which puts the reasoning into perspective. In any case, the capacitor with the lowest self-inductance should be placed closest to the supply pin.

Some Additional Notes about Decoupling Capacitors

For microcontrollers and FPGAs, a recommendation for the number and type of supply decoupling capacitors is sometimes given in the datasheet. The manufacturer often recommends more than enough decoupling capacitors to be on the safe side. At the end of the day, you are the one who has to deal with the cost and space involved. Therefore, I usually do not follow the suggestion right away; I usually start with half the recommended number. In the layout, however, I allow for the possibility of assembling most of them myself to be on the safe side.

Did you know that supply decoupling capacitors are built into every IC? This is achieved by using the gate-source capacitance of otherwise unused MOSFETs. Such decoupling capacitors can reach hundreds of picofarads. Adding supply decoupling capacitors outside the IC is only necessary because larger capacitance values would take up too much die space, about 1 mm^2 of die area per 5 nF [R072]. The static memory ICs are an exception. They have up to 20 nF of capacitance formed from non-switching cells [R121]. Fortunately, the inductance of the bonding wires is sufficiently small that power supply decoupling outside the IC package is effective from about 1 nF and up for conventional ICs.

So far, decoupling capacitors have been considered and used from the perspective of preventing supply voltage dips. However, the solution can also be seen from the "other side". Without a decoupling capacitor, there are rapidly increasing currents throughout the supply network, which can lead to electromagnetic interference (EMI) problems due to magnetic fields.

98 Direct-Current Supply Design

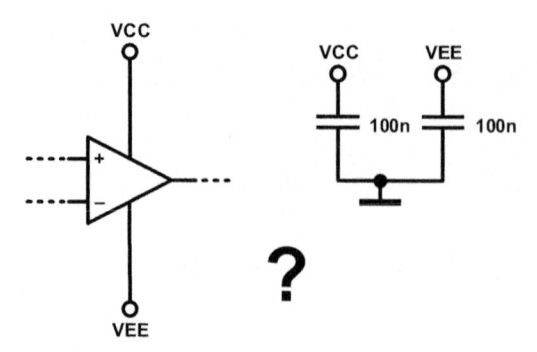

Figure 46. Operational amplifier and its decoupling capacitors on a schematic.

Analog ICs typically operate in a limited frequency band—in extreme cases, only at one frequency. A decoupling capacitor helps keep the narrow-band current draw of the IC away from the power supply. As a rule of thumb, a decoupling capacitor is recommended on the supply pins of an analog device that processes signals of 100 kHz or more. The capacitance depends on the power supply cleanliness requirements.

Many things that are not visible in the schematic play an important role in the layout. In Figure 46, the decoupling capacitors for the operational amplifier are drawn separately from the amplifier on the schematic sheet. This is a permissible means of making the functional part of the circuit clearer. However, the problem with this is that when a third party designs the layout, the decoupling capacitors may end up in the wrong place. One solution is to leave a note in the schematic for the layout designer, as shown in Figure 47.

Autorouters sometimes allow "shared vias" by default. If you do not disable this option, the autorouter may create long ground connections. This is especially the case if the ground connections of two or more capacitors are combined on one via (see Figure 48 for examples). Long ground lines are prone to significant voltage drops caused by trace inductance. This ground bounce raises the potential and reduces the noise margin at the output of an affected digital IC. Decoupling capacitors cannot be used to suppress this ground noise.

It is better to disable the shared via option and provide each IC with an exclusive via to the ground plane. This is the closest approximation

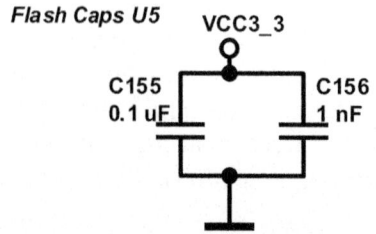

Figure 47. Note about the location of the decoupling capacitors.

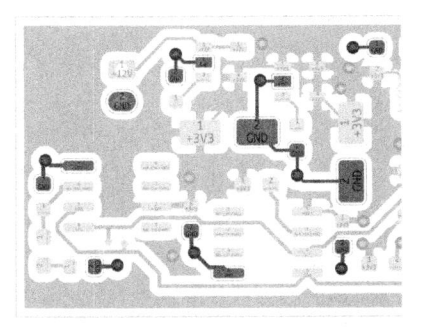

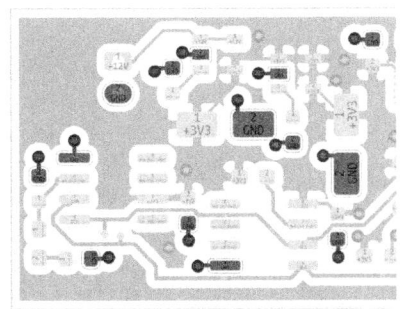

Figure 48. Left: Partially long leads to ground through common vias. Right: Short ground traces through individual vias.

to star grounding, where all components see exactly the same ground potential. It is still an approximation due to the low but not vanishing inductance of the ground plane.

In a dense layout, shared vias are sometimes tempting. When merging two existing vias in the layout to save space, the layout designer must always be aware of what he is merging and whether it is allowed. Shared vias of input and output stages must be critically evaluated.

Solutions When Decoupling Caps Are Not Enough

What can be done if far too many decoupling capacitors would be required to keep the voltage drop within specifications? One approach is to switch to true differential logic, such as LVDS. Instead of taking charge from the source each time a bit is toggled and returning it to the source the next time it is toggled, differential connections essentially just move charge from one transistor gate capacitor to the other. Note that this is only true for true differential systems. USB, for example, is not true differential: the two data lines are each driven by a voltage source. Although they operate in phase opposition, they draw their charge from the power supply.

Alternatively, the board can be operated from a system voltage of 12 V or 24 V. A system voltage of 48 V is also common, but is disproportionately more expensive than the lower voltages, both in terms of components and testing. Where necessary, the digital supply voltages are generated locally using DC/DC converters.

Chokes in the Supply Path

A pi filter was recommended with the DC/DC converter, which includes a choke. In a reduced form, as an apparent second-order filter, the combination of a choke and a decoupling capacitor is sometimes seen in schematics, as shown in Figure 49.

A few comments about this filter are as follows:

100 Direct-Current Supply Design

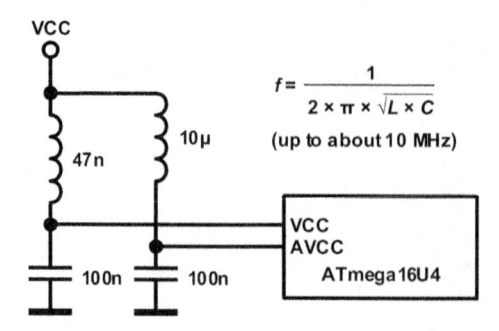

Figure 49. Chokes in the supply path.

- In the chapter on robust interfaces, the choice of choke type is discussed in more detail, and it is also shown that the choke is effective not only in combination with the capacitor, but also on its own due to its core losses. This means, however, that it cannot be selected solely on the basis of its inductance.

- As a rule of thumb, a choke is purely inductive up to about 10 MHz, and the formula for the cutoff frequency is relevant; for example, 160 kHz for AVCC and 2.3 MHz for VCC in Figure 48.

- If you choose the choke in a standard package—size 0805, 0603, or 0402—you can install a 0 Ω bridge at the outset to save cost, and use the choke only when necessary.

Multiple Ground Planes, or one Large Ground Plane?

Analog-digital mixed-signal circuits raise the question of whether the ground plane should be divided into distinct regions called "localized reference planes". If you decide to use separate ground planes (e.g., an analog ground and a digital ground), the following steps are recommended:

- Select a minimum horizontal distance between the planes, a minimum moat width of 3 mm is recommended.

- No copper in the moat. Check the PCBs against the light.

- To avoid too-large voltage differences between the otherwise isolated ground planes in the case of ESD, bridge the moat with Schottky diodes or with a choke. Schottky diodes respond quickly to an ESD event, but they are a path for high-frequency vagrant currents, possibly canceling the split ground plane advantages. Using a choke, the parasitic capacitance poses a path for high-frequency vagrant currents and may cancel the localized plane concept too.

The following are some instructions for signal routing across a moat:

- The safest way to cross the moat is to use optocouplers or transformers.

- Single high-frequency signals can be routed over the moat if the moat is bridged with a capacitor on the ground plane side and the power plane, if present, is locally decoupled.

- Differential signals can be routed directly across the moat. Common-mode chokes on either side prevent stray currents.

- An ADC across the moat is allowed only if the analog and digital sides of the ADC are galvanically isolated.

When analog and digital signals leave a board with localized reference planes, the return current should flow back to its own ground connection (i.e., analog ground or digital ground). This is only guaranteed if the planes are not connected somewhere. However, this leads to the problem of high ESD sensitivity as discussed above. Therefore, local reference planes are difficult to implement in this case.

A modification of the localized reference plane concept allows for direct but limited electrical connection in the planes, as shown below in Figure 50:

- For cross border connections, the signals are bundled and a bridge is created between the ground planes underneath the traces.

- No signal or power trace may cross the moat outside the bridge.

- If a power trace crosses the bridge (e.g., if an analog supply voltage is made from a digital supply), at least one choke must be used in between, adding a voltage regulator would be best.

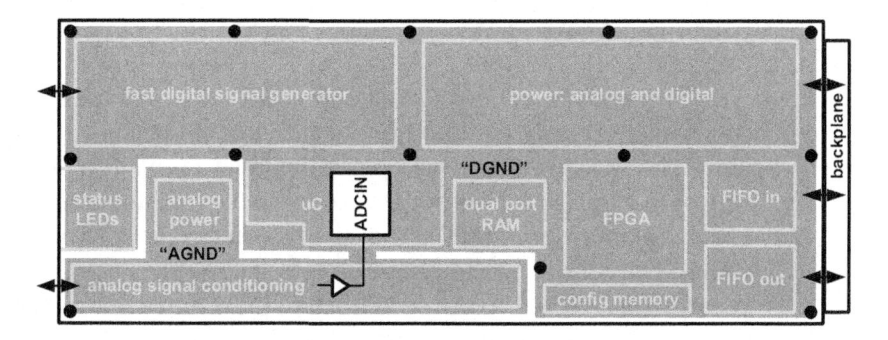

Figure 50. Dedicated ground plane for analog and digital domains.

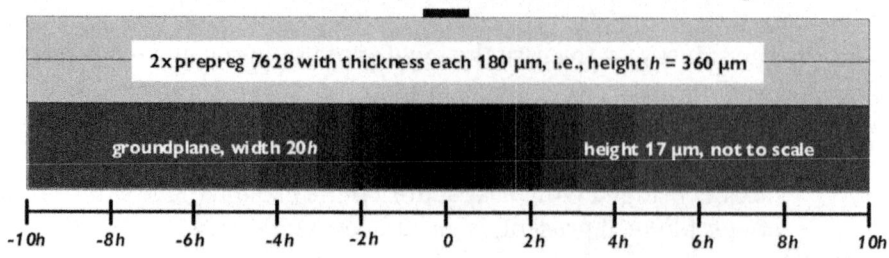

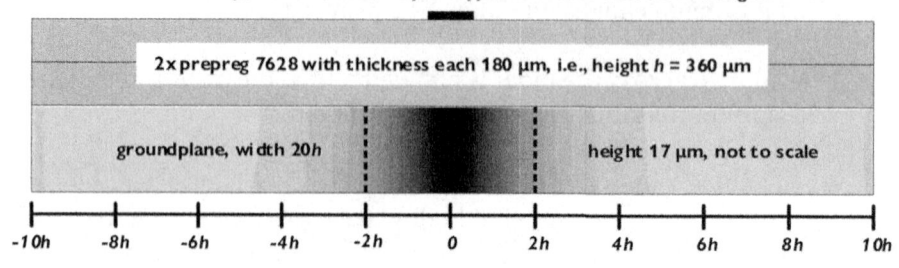

Figure 51. Current distribution in a ground plane under a trace. Result based on a simulation with FEMM [193] at 100 kHz and 1 MHz. An analysis shows that the area bounded by the dashed lines contains 36% of the total current at 100 kHz and 71% at 1 MHz. The current concentration in the return path at 1 MHz is based on the proximity effect. From about 10 MHz, there is even an additional concentration due to the skin effect.

When using multiple ADCs, the bridge quickly becomes such a wide connection that you are effectively back to a single large ground plane. In fact, you can deliberately use a single ground plane across the entire board, even for mixed-signal circuits. Here are the requirements:

- Clean partition. Define the digital domain and the analog domain. Under no circumstances should signals from one area be routed into the other. An "invisible moat" (i.e., a keep-out area in the layout) helps to keep the partition clean.

- Partition boundaries that are large enough (e.g., 1 cm wide).

- A large ground plane in a mixed-signal circuit works because fast return currents are confined to the area under the forward trace when the distance to the next potential layer is small. When the distance from the center of the trace is twice the height above the ground plane, more than 70% of the return current of a 1 MHz AC signal is contained within that distance (see Figure 51). This is sometimes referred to as the "2H rule". Choosing a boundary many times wider than the distance between the signal plane and the ground plane will prevent high frequency stray return currents from flowing back into the wrong ground plane area.

Ground Plane to Chassis Connection

Our discussion of connecting the ground plane to the conductive enclosure is limited to devices with a low DC voltage supply. For equipment with mains supply or high DC voltages, the paths for residual and leakage currents must be clarified in accordance with the standards; it is best to contact a test center for this, as it can be a rather complex task. In this discussion, the main issue is how to keep electrostatic discharge currents out of the device.

There is no question that the ground plane must be connected to a conductive case because that is the great advantage of such a case: electrostatic discharges reaching the PCB can be conducted directly to the case via a conductive connection. They do not remain somewhere, accumulate and lead to destructive discharges on the PCB, as is the case with a PCB in a plastic, non-conductive enclosure. In the latter case, contact an EMC test center as soon as possible, as it is difficult to get a device without a conductive case through all the tests.

Obviously, a connection to the case must be used wherever an electrostatic discharge can reach the board: for example, at the location of connectors, display and ventilation openings. The connection should be as low impedance as possible so that all the discharge current leaves the board immediately. This has several consequences:

- A conductive case acts like an antenna or electrode, picking up radiation and interference fields from the environment. By connecting directly to the PCB ground, these fields reach the circuit board. For this reason, it is important to place capacitors directly from the ground plane to the power plane(s). A 1 nF capacitor in parallel with a 100 nF capacitor is recommended near each plated hole used for a connecting screw. As a result, the power plane(s) will experience the same potential fluctuations as the ground plane, making them invisible to the circuit.

- In the case of multiple ground planes, the process of connecting more than one of them with as low impedance as possible to the conductive enclosure will destroy their separation. This is one of the reasons for using a single large ground plane for the entire board. However, there are situations where multiple ground planes can be kept separate. One example is an analog part on the PCB connected to a fully shielded sensor. The sensor's shield is connected to the housing, making the sensor an extension of the housing. There is no need to connect the analog ground at the sensor connector location to the enclosure.

- Multiple ground plane(s), power plane(s), and chassis connections introduce paths for roaming high frequency currents. A single PCB-to-chassis connection—a star ground—would prevent this. If your device has only one connector and no other openings in the case, such a single connection would automatically follow the advice just given. In reality, however, most devices are more complex and have many openings and connectors. Omitting the PCB-to-chassis connection for all but one connector essentially means leaving doors open for ESD discharge.

- Some engineers argue for even more PCB-to-chassis interconnects than we have discussed so far. They recommend to place extra connections spaced at least $\lambda/20$ apart where $\lambda = c_0 \times$ rise time$/(0.32 \times \varepsilon_r)$ and c_0 is the speed of light, the rise time of the fastest digital signal on the board and the dielectric constant ε_r of the layer(s) between the power plane and ground planes. The reasoning behind this measure is that the capacitors placed between the power plane(s) and the ground plane(s) only connect them locally in phase. Further away, the equalization diverges due to the propagation time of very fast signals on the PCB. This principle is called "controlled multipoint grounding". The many black dots in Figure 50 above, each representing a PCB-to-chassis connection, are an example of an implementation of this principle.

Multiple Power Planes

In the section on power supply decoupling above, it was shown that it is advantageous to provide a power plane because it connects all decoupling capacitors with low inductance. Ideally, therefore, a power plane should be implemented for each supply voltage, ideally extending over the entire PCB area. If there are not enough layers available, you can try to create a power plane by filling layers. See the section about filling layers above.

Finally, the last option is to split a layer or flood a layer in different areas to realize power planes of different voltages in one layer. If there are multiple coplanar power planes, the 20H rule must be followed. As mentioned above, but applied to the situation here, the 20H rule states that there should be a distance of $20 \times H$ between two edges of different coplanar power planes, where H is the height of the power planes above the ground plane, see Figure 52. This is based on the fact that approximately 70% of the stray field is then contained within the gap. This is considered a reasonable compromise. At $100 \times H$, it would be 98%, but also a correspondingly large reduction in power plane size.

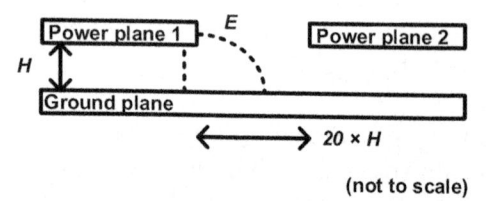

Figure 52. Example implementation of the 20H rule.

3 Robust Interfaces

Interfaces always have intended parts, but they also have unintended parts. For the hardware designer, the intended parts are the communication standards used (e.g., CMOS native or RS-485). The unwanted parts are electromagnetic compliance (EMC) problems, namely inadequate immunity to electrostatic discharge (ESD), fast electrical transients (EFT, burst), slow high-energy overvoltages (surge), interference due to coupling/uncoupling of fields, and input/output of electromagnetic waves. As a result, this chapter is organized as follows:

- Comments on communications standards at the hardware level.

- Safety of the interface—i.e. protection against ESD, surge, burst and ground potential differences.

- Interface integrity and susceptibility to interference.

- Examples of interfaces for selected communications standards.

- Notes on EMC testing.

Notes on the I^{2}C Interface

At first glance, the I^{2}C interface appears to be an elegant communication standard. With only two PCB tracks, 127 ICs can be connected and controlled. However, anyone who has ever worked with I^{2}C knows its drawbacks. ICs with an I^{2}C interface usually have a very limited address space. Conflicts quickly arise. One solution is to add a switch to the data track. This allows you to control two ICs with identical I^{2}C addresses. Alternatively, use a second I^{2}C interface altogether; many microcontrollers provide more than one I^{2}C port. Using such a workaround eliminates the advantage of the compact I^{2}C interface over the multiline SPI interface.

There are other drawbacks. Certain ICs with an I^{2}C interface require clock stretching, where the serial clock (SCL) is pulled to ground by the slave until it is ready to respond. Clock stretching is part of the I^{2}C specification, but it is not mandatory to implement it. This capability is not guaranteed for any particular microcontroller or peripheral IC. There is no timeout for clock stretching. A defective or incorrectly programmed IC on the I^{2}C bus can disable it forever.

I²C has two speed modes: 100 kbps and 400 kbps. Not every IC supports the faster mode. The faster mode may also be unattainable due to too long traces or too much parasitic capacitance due to too many receivers. Since the HIGH level is not driven and the pull-up resistor forms an RC circuit with the input capacitance, the HIGH level is reached with a corresponding delay. According to the standard, 400 pF is the maximum load capacitance on an I²C line at 100 kbps. Since the data line is bi-directional, a normal driver cannot simply be inserted in between; a special I²C bus extender IC is required.

Fast mode can quickly become a bottleneck for devices that transmit more than a few bits. Most of the time these devices do not have an I²C interface at all, but it can happen. With EEPROMS and displays, you have to check carefully if the speed is sufficient. It is not surprising that there are no flash memories with an I²C interface, it would simply be too slow.

If the transmitter and receiver have different supply voltages, translation is required. The simplest is a longitudinal FET [R189].

Not all ICs with I²C interface are multi-master capable. This can be a real problem if you have not checked this before. If, for example, you depend on interrupts by the slaves, all ICs must have arbitration in case of conflicts of simultaneous access. If this is not the case, you must split the bus into two, provided you have two I²C ports available at the microcontroller; otherwise, the only option is to replace the peripheral IC, which is not multi-master capable.

Finally, it can be stated that with the I²C interface, almost like nowhere else, it is fundamental to thoroughly examine the data traffic with a logic analyzer. The complexity of addressing by means of addresses and confirmations of transmissions often necessitates investigation at the lowest level. This means that a new I²C connection takes time to set up, and, as a rule, it does not work at the first attempt, certainly not, if implemented blindly.

Notes on the SPI Interface

An SPI connection is much easier to set up than an I²C connection, although it requires more tracks and a select chip per slave for multiple participants. If there are many peripheral ICs, a 3-to-8 decoder IC 238 can be used to reduce the number of microcontroller pins involved. SPI knows only one master on the bus. If a slave alarms the microcontroller, an additional interrupt line is required (i.e. there are no conflicts). There are no predefined flow controls or acknowledgements. All lines can easily be equipped with drivers, and different voltage levels can be addressed. There is no maximum speed defined for SPI, and the clock rate can be irregular. Communication at e.g. 20 Mbps is quite possible, but the datasheets of all ICs involved must be checked carefully for their maximum speed.

In general, SPI does not need pull-up resistors because all lines are always driven. However, if there is a possibility that no single slave is connected, it is preferable to place a 10 kΩ pull-up resistor on MISO. Otherwise, the undefined potential can lead to increased current consumption due to a CMOS shoot-through current.

With a Flash card, not only the MISO line must have a pull-up resistor, but preferably all lines. The Flash memory specification says that all lines must have one, but this is because the pins are natively designed to be open-collector. In SPI mode, the board drives the MISO pin. Nevertheless, you should put pull-up resistors everywhere, because until the card is in SPI mode, you will have an increased current draw. You can save a few milliamperes if you put a pull-up resistor on all unused pins of a Flash card. Another thing to note is that there is no guarantee that the Flash card will set the MISO pin to a high impedance state when the chip select goes high again. One advantage is that there is no other slave to put on the bus when driving a Flash card.

Low Voltage Differential Signaling Explained

With strong and fast drivers, you can theoretically operate at very high SPI data rates. Reflections would have to be expected, as explained in the chapter about signals on PCBs, but due to the unidirectional nature of the lines, you could easily counteract them with a suitable series or parallel terminating resistor. Nevertheless, no current high-speed connection uses the SPI standard. The reason for this is that in single-ended mode, the individual bits are transmitted asymmetrically: only the corresponding line is driven, not a return line. Asymmetric drivers cause the following problems:

- Each bit change from logical 0 to 1 draws current from the source. Decoupling capacitors can keep the supply clean to a certain extent, but at very fast communication speeds, this cleaning is not effective enough (see the chapter on supply design). The noise on the supply appears one-to-one on the data signal and reduces the safe margin to the logic undefined state.

- Each time a bit changes from a logical 1 to a logical 0, a current is injected into the ground, causing increased noise in the ground plane. This noise is in turn found on the data lines of all the digital ICs in the vicinity. There is no antidote; decoupling capacitors do not help in this case.

- Using a cable between master and slave, twisting the supply line with the ground line helps only in the 0 to 1 case, in the 1 to 0 case the data line should be twisted with the ground line. It is not possible to do both at the same time, which means that EMC tests may fail at high-speed emissions.

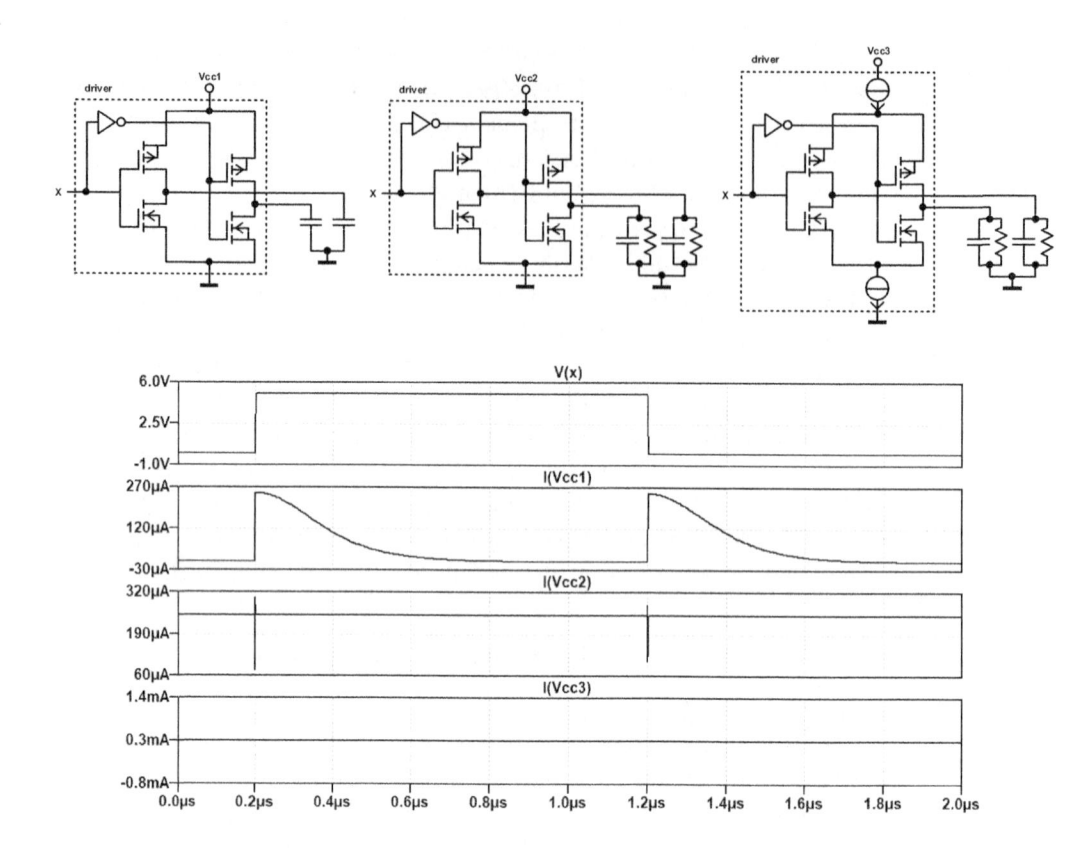

Figure 53. Top: From a counter-clocked CMOS output stage to a true LVDS system. Bottom: Voltage curve of the drive signal X and corresponding currents in the supply lines for all three circuits.

For a differential signal, there will be simultaneous emissions with a 0° phase shift and a 180° phase shift. The radiation patterns of the two traces may not be the same, and they are separated by a small distance; however, the total radiation is small when viewed from the far field. Sufficiently low-emission transmission of very fast signals is only possible with a differential scheme. However, differential transmission alone may not be sufficient. This is explained in Figure 53. First, differential transmission can be achieved by simply switching two CMOS outputs in opposite directions. However, this results in current being drawn from the source on each signal edge, as one of the two PMOS outputs switches alternately. Adding parallel terminations results in a constant current draw, except for peaks at the edge times. Inserting a current source into the driver at the voltage input and before the ground output provides the desired solution. Now the driver stage draws a constant current without any peaks. Although the signals on the lines are already considered differential with the counter-clocked CMOS output stage, only the solution with the current sources leads to a truly differential driver, a driver whose current consumption is independent of the data. Low voltage differential signaling (LVDS) uses this scheme.

Purely because of the need to keep the supply clean, LVDS has become the only solution for very fast signal transmission in the gigabit/s range. However, it is also advantageous at lower frequencies because the signal swing is typically only 340 mV. The swing is determined by the 3.4 mA continuous current flowing in either direction through a 100 Ω termination resistor at 50 Ω line impedance. This results in lower losses than many other communication standards.

RS-422, RS-485, CAN, and USB

The RS-422 and RS-485 standards have a differential signal scheme, but the drivers themselves are not strictly differential, as explained above. Assuming a minimum signal swing of 400 mV, they are equivalent to LVDS from an EMC point of view, if we consider only the transmission lines. On the PCB itself, they generate noise with every switch.

The same can be said for the CAN bus, with the difference that the voltage swing is much higher by default, in order to transmit signals as safely as possible in an environment with a high density of EMI without adding too much to the interference level itself.

All of the differential transmission standards mentioned so far serve a clear application segment. LVDS is the choice for point-to-point gigabit/s connections, but limited to a distance of about 10 m. The most prominent application is Serial ATA. RS-422 and RS-485 are the choices for bridging distances up to 1.2 km (i.e., for low-noise data lines in industrial installations with high common-mode rejection). CAN, with its built-in data protocol that requires a five-fold error check of each bit, is the standard for very secure data transmission in highly disturbed environments, such as vehicles with spark plugs.

Microcontrollers with built-in USB ports are now available. The most important aspect of a USB port from a hardware engineer's point of view is its ESD protection. USB connections are typically made and released very often, usually by hand, which increases the risk of ESD.

Electrostatic Discharge Protection

The discussion of practical ESD aspects begins with the example shown in Figure 54. In this case, the KMA199E angle sensor with digital inter-

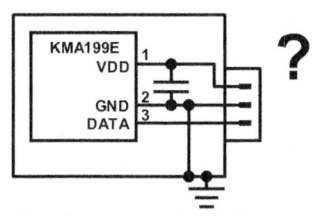

Figure 54. An example to introduce the topic of ESD testing: KMA199E angle sensor (automotive) with digital interface and 8 kV ESD resistance according to HBM. The sensor is equipped with a connector for assembly and repair and is mounted in a metal box.

face is fitted with a connector and installed in a metal enclosure for assembly and repair purposes.

The developer of the circuit reported the following problem: "During the ESD tests in the EMC test center, all the sensors broke, although the sensor can withstand 8 kV according to the data sheet!"

The reason for the failure is the keyword "HBM" in the data sheet of the component, which means the test case "Human Body Model". Today this is an ESD component test; it used to be a system test in the USA, also called MIL-STD-883. However, the entire device, in the case of Figure 54, consisting of the sensor and a connector, is subjected to an ESD system test according to IEC 61000-4-2 with much harsher conditions in the EMC test center, hence the defects.

The HBM test confirms protection against ESD damage in an environment with targeted low ESD levels, such as the development environment with ESD mat and wrist strap, ESD shoes, dissipative floor, and the production environment with these measures and controlled humidity.

Instead, the IEC 61000-4-2 system test simulates ESD levels that may occur in practice when the device is used in the field. In most cases, it is not possible to assume an environment where the ESD hazard is deliberately kept low. On the contrary, the worst-case scenario of an extremely hostile ESD environment must be assumed. Therefore, the system test is much more stringent than the component test.

Table 21 quantifies the differences between the two ESD test schemes. The test according to IEC 61000-4-2 is associated with a pulse energy that is about four times higher than in the HBM case. The maximum voltage is reached after 1 ns and about half of the energy is converted after 25 ns compared to the HBM test. In the HBM test, the rise time of 25 ns is much longer, combined with a comparatively "leisurely" energy transfer.

From a technical point of view, it is no problem to implement protection according to IEC 61000-4-2 with 8 kV pulses on the IC die. For example, the ADN4666 LVDS line receiver or the MAX14840 RS-485 half-duplex transceiver have such protection, the latter even for 12 kV in the SO8 package. These ICs are most likely to be connected directly to a connector. However, implementing IEC 61000-4-2 protection in an IC itself requires a lot of board space. This would unnecessarily increase the price of many components that normally have no direct contact with the world outside the PCB [R034, R035].

However, semiconductor manufacturers do not want to rely solely on ESD protection measures in development and production. They know all too well that these protective measures—everything dissipative and electrically connected, from chairs to shoes to table tops to soldering irons—are too often neglected. While this is a sin of omission on the part of the user, the mass of defective components automatically reflects poorly on the component manufacturer. The logical compromise is to incorporate reduced ESD protection suitable for the design

Variable	Human body model	IEC 61000-4-2
Usability	Human-to-IC discharges in environments that only partially allow electrostatic charges*	Worst case assumption of a completely ESD friendly environment
Generator capacity	100 pF	150 pF
Discharge resistor	1500 Ω	330 Ω
Current peak	5.33 A at 8 kV contact voltage	24.2 A at 8 kV contact voltage
Energy	1.5 mJ	6 mJ
Waveform		
Pulse rise time	25 ns	< 1 ns
Minimum pulses per pin	1 x positive, 1 x negative	10 x positive, 10 x negative

* e.g., Table conductive, but shoes not.

Table 21. Comparison of HBM vs. IEC 61000-4-2 testing, according to [R031].

and manufacturing environment. Manufacturers are helped by the fact that they can achieve limited, but existing, partial ESD protection with the standard PN diodes of the IC design process by switching one diode in the reverse direction to ground and one diode in the forward direction to the supply voltage at each input. The diode to the supply voltage is necessary for positive overvoltages because a Zener diode would be too large and therefore too expensive. This means that positive voltage spikes are diverted to the power supply instead of to ground. The protection provided by standard semiconductor PN diodes is limited because these diodes are too small to repeatedly absorb the full ESD current specified in IEC 61000-4-2 without overheating. In addition, a high ESD current through the chip diode to ground will cause the forward voltage to rise above the supply voltage and damage the IC [R034].

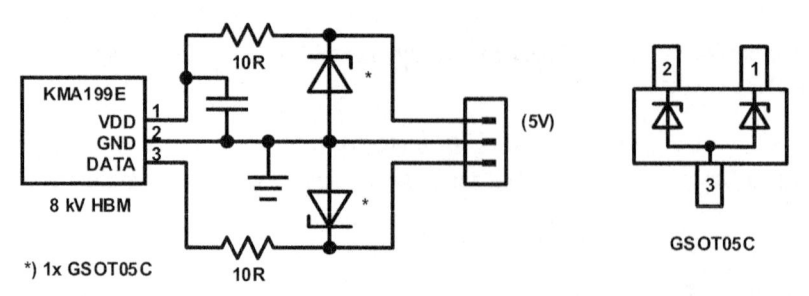

Figure 55. ESD protection consisting of an avalanche diode and an optional resistor, see text for details. The GSOT5C protects two 5V lines simultaneously.

ESD protection of interface pins is typically achieved using avalanche diodes. Figure 55 shows the implementation using the GSOT05C IC for our introductory example. Avalanche diodes are diodes whose avalanche breakdown voltage has been placed on the Zener breakdown voltage by doping, so that an abrupt breakdown occurs. Furthermore, the diodes are designed in such a way that they can carry high currents for a short time, even during the breakdown, currents at which ordinary diodes would have failed by now. Avalanche diodes are explicit protection devices that are not normally used for other purposes. They are known by various names, some of which are brand names: Transient Voltage Suppressor Diode (TVS), Transzorb™ (Vishay Intertechnology, Inc.), Transil™ (ST Microelectronics).

Why are there resistors connected to the diodes in Figure 55? In fact, this resistor is only mentioned in a few publications [R035, R044]. Figure 56 can be used to explain what happens without resistors. Due to the characteristics of the GSOT05C avalanche diodes, a residual voltage of 16 V can remain across the diode in the event of ESD. Since the data input is internally protected against VDD by a PN diode, albeit only for the HBM level, a current is generated through the discharged decoupling capacitor. The magnitude of this current can be high, since it is only determined by the line resistances and the ESR of the capacitor. Damage is possible.

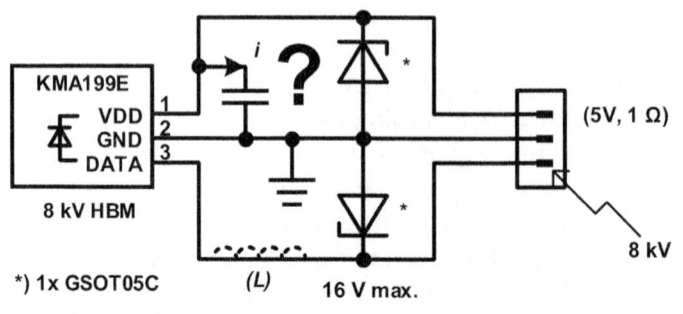

Figure 56. ESD protection consisting of an avalanche diode without resistor.

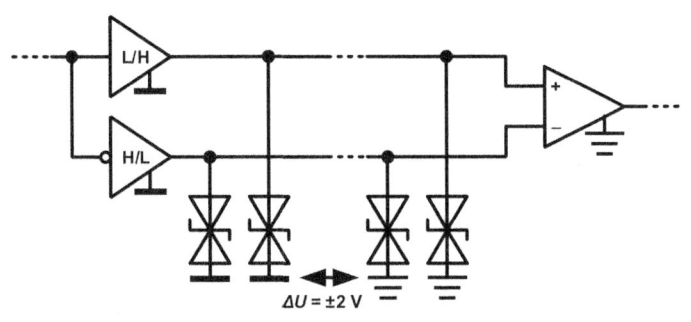

Figure 57. ESD protection for connections with negative voltages, example of a CAN connection (from R035).

The 10 Ω resistor limits this current to a maximum of 160 mA, or even less due to the forward voltage of the PN diode. In these calculations, the PCB trace inductance was neglected because the distances were short. If the IC were not so close to the protection diodes, the trace inductance would prevent a steep rise in current, and resistors may not be necessary.

The solution shown in Figure 55 allows only unipolar positive signals; negative voltages are conducted to ground in the forward direction of the diodes. However, by using two reverse-polarity avalanche diodes, available in a single package as a "bidirectional avalanche diode," bipolar data lines with positive and negative voltages can also be protected. Figure 57 shows an example of ESD protection for a CAN bus that has normal signal levels of 0 V/5 V, but a permitted common-mode voltage shift of ±2 V.

Types SMAJXA/CA and SMBJXA/CA are often used as protective diodes, where X represents the operating voltage (e.g., SMBJ5 for 5 V operating voltage), the suffix "A" indicates unidirectional protection, and the suffix "CA" indicates bidirectional protection. The suffix "SMA" or "SMB" refers to the package shape. Part "J" defines the device as an avalanche diode. SMAJ and SMBJ diodes are produced by several manufacturers. However, they are oversized for ESD protection only. SMAJ and SMBJ diodes also protect against IEC 61000-4-4 burst events and IEC 61000-4-5 surge events, but more on this later.

However, SMAJ and SMBJ diodes are often used for pure ESD protection because they are more readily available than pure ESD protection diodes.

Table 22 (on the next page) provides a general comparison between the two, considering multiple diodes and diode arrays.

Of particular note is the difference in junction capacitance: a GSOT5C diode has 350 pF and an SMAJ5.0 has 2 nF at 1 MHz.

Protective Elements	Advantages	Disadvantages
Dedicated ESD protection diode, (e.g., GSOT[1], ESDALC[2]).	Space saving. Small capacity.	ESD protection only. Few pin-compatible alternatives.
General TVS diode, (e.g., SMAJ, SMBJ).	Multiple manufacturers. Protects against IEC 61000-4-4 ("Burst") and IEC 61000-4-5 ("Surge").	Relatively large.
Protective diode array.	Space saving. Compact layout.	Few to no pin-compatible alternatives.

[1] Vishay Intertechnology, Inc. [2] ST Microelectronics.

Table 22. Overview of protection diode selection.

Variable	Pulse rise time t_r limitation (10% to 90% level rise time)
Cut-off frequency of a low pass	$f_g = 1/(2 \times \pi \times R \times C)$
Step signal rise time	$t_r \approx 0.35/f_g$
Maximum symbol rate	at 90% modulation: 1 bit/$(2 \times t_r)$
Example with termination resistor $R = 27\ \Omega$	GSOT5C: $C = 350$ pF → 24.1 Mbit/s SMAJA5.0: $C = 2$ nF → 4.2 Mbit/s
USB 1.0	Low Speed: 1.5 Mbit/s, High-Speed 12 Mbit/s
High Speed USB 2.0	480 Mbit/s
Firewire 400 (LVDS)	400 Mbit/s

Table 23. Avalanche diode data rate limiting.

ESD Protection for High-Speed Interfaces

Table 23 shows the maximum data rates possible with a GSOT5C or SMAJ5.0 diode driven from a common CMOS source. Fast data transfer requires low junction capacitance. The special structure shown in Figure 58 with so-called hiding diodes achieves this. These hiding diodes around the avalanche diode are normal PN diodes but trimmed for low junction capacitance. Connecting the upper PN diodes to VBUS in Figure 58 provides this junction capacitance. In series with the avalanche diode, the total capacitance to ground is even less than the PN capacitance. For example, in the case of the TPD2E001 in Figure 58, the load capacitance on the line is only 1.5 pF. The data sheet specifies a data rate of 240 Mbps for signals on this line.

In the circuit shown in Figure 58, the power supply is also protected from an ESD event with the 100 nF capacitor absorbing some of the energy. The capacitor is required in this situation to meet the ESD rating.

The structure shown in Figure 58 is obviously only suitable for unipolar signals. However, there are devices on the market that work without a DC voltage for the PN diodes (i.e., do not require the VBUS voltage shown in Figure 58). For example, the TPD2E2U06 device has a load capacitance of only 1.5 pF, even without a VBUS connection.

Some layout notes regarding ESD diodes [R036]:

- An ESD protection diode must be placed as close as possible to the connector.

- The diode should be connected directly to a ground plane. The via must not be shared.

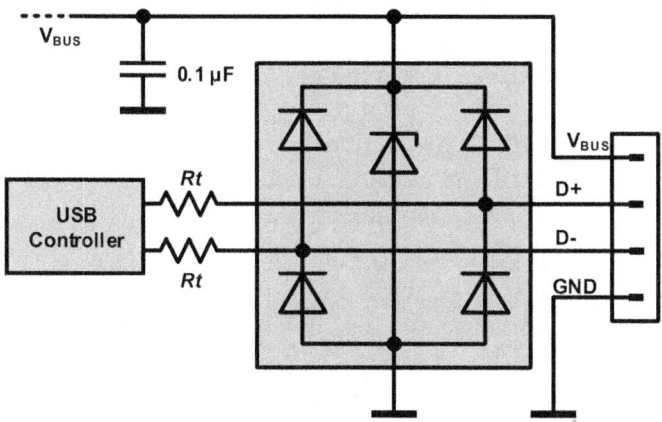

Figure 58. ESD protection for high-speed data links. A positive pulse on D+ is passed without delay through the PN diode on the top right of the avalanche diode, which discharges the overvoltage to ground. A negative pulse on D+ goes directly to ground through the PN diode on the bottom right.

- The ground plane should be connected to the chassis with low impedance at the connector; however, if the chassis absorbs a lot of noise, a 4.7 nF, 250 VAC capacitor in parallel with a 1 MΩ resistor can be used for this connection [R040].

- Unprotected data lines should not be routed in parallel with protected lines.

- Refer to the data sheet layout and application note(s).

ESD Protection for Switches and Keypads

When we think of ESD, we often think of connector interfaces. However, switches, buttons, keypads, and the like are often more exposed to ESD than connectors, and can even be ESD sources themselves due to repeated mechanical actuation and triboelectricity. Therefore, the connections to these elements should also be protected from ESD. See for example Figure 59.

ESD Protection for Instrumentation Amplifiers

A small current leakage in the protection diodes is required not only for fast signals. Instrumentation amplifiers (INAs) with input resistances in the GΩ range cannot be protected with SMBJ5.0CA diodes:

- The leakage current of the SMBJ5.0CA diode cancels out the high input resistance of the instrumentation amplifier, reducing it to approximately 10 kΩ, since the diode has a leakage current of 35 µA at 300 mV.

- Due to the different leakage currents in the diodes and the different source impedances of the signals at the INA inputs, the common mode rejection is degraded and may not be sufficient.

- Even with a GSOT05C protective diode, there would be a greatly reduced input resistance of about 600 kΩ, far from 10 GΩ, with correspondingly poor common-mode rejection.

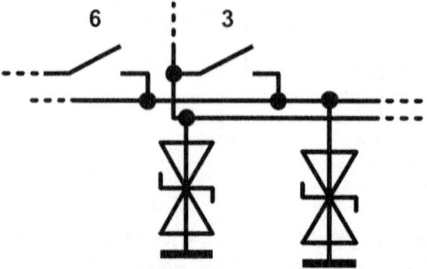

Figure 59. ESD protection for a numeric keypad, detail.

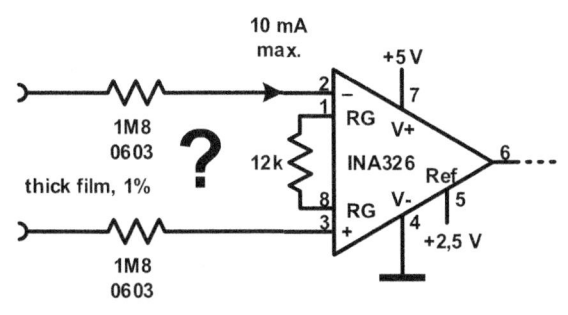

Figure 60. Attempt to protect an instrumentation amplifier from an ESD event using series thick-film resistors.

Therefore, we must use a different protection solution. In the example in Figure 60, an attempt was made to use series resistors to protect the input of the instrumentation amplifier. Since there are no recommendations in the INA data sheet, the value of the resistor was calculated as follows:

- 8 kV divided by 10 mA—the "absolute maximum input current" according to the INA data sheet—gives a value of 800 kΩ.

- 100% derating results in 1.6 MΩ, next is a higher E12 series value of 1.8 MΩ.

A standard 1% thick film resistor was used. However, due to the very high input impedance of the instrumentation amplifier in the range of 10 GΩ, even a larger tolerance of the protective resistor would have no effect on the common mode rejection.

The question mark in Figure 60 refers to the problem that an 0603 thick-film resistor is not 8 kV ESD withstand. A thin-film type would be even less ESD resistant. Thin-film resistors consist of a layer of metal that is only a few hundred angstroms thick. This severely limits the device's ability to withstand the energy that passes through it during an electrostatic discharge, making it very susceptible to ESD damage (R182). The resistor type must be carbon composite or MELF, both of which are immune to ESD events [R045, R046]. Carbon composite resistors are still readily available, especially for high-voltage pulse withstand purposes, but practically only in leaded versions. MELF resistors consist of a metal oxide layer sputtered onto a cylindrical ceramic substrate. Due to the much larger surface area than a conventional chip resistor, pulse energy can be better dissipated.

The criterion for the impulse resistance of a resistor is usually that the change in value does not leave the tolerance range. For the protection application of the standard solution, much larger changes would actually be permissible (e.g. 10%). But there is no data available, not even from the manufacturer. I did some tests myself and showed that larger resistance changes in the 10% range are usually associated with a much larger change in the 50% range at the next ESD pulse. This means that as soon as the resistance value is out of tolerance, you can expect a serious change in the following ESD pulses. Therefore, the limits set by the manufacturer are valid, even if larger resistance changes are acceptable.

ESD Protection Using a Capacitor

ESD protection by means of a capacitor is widely used in consumer electronics and automotive engineering due to its lower cost compared to a diode [R048]. At the other end of the price scale are connectors with integrated ESD capacitors, which are much more expensive than a conventional connector plus avalanche diodes. They are mainly used in military applications, where they also protect against nuclear electromagnetic pulses.

The basis of the protection is a capacitive voltage divider. The ESD gun has an internal 150 pF capacitor. Using a 1.5 nF capacitor with a sufficiently high voltage rating, 800 V of an 8 kV pulse remains to be absorbed by the IC's internal ESD protection diodes. However, this calculation only works if the capacitor is placed as close as possible to the connector, otherwise a resonant circuit with the track inductance is created, resulting in an even higher peak voltage. Because of this danger, some explicitly advise against using capacitors to protect against ESD [R049].

ESD Protection Using a Varistor

At high voltages, protective diodes may no longer be possible, in which case varistors are used. Varistors are resistors whose resistance stops above a certain voltage. Varistors have a slightly longer response time than avalanche diodes. They also have a limited number of low pulse counts, such as five pulses, before degenerating. In addition, they exhibit significant aging behavior and large leakage current. If possible, avalanche diodes are clearly the better solution [R039].

Plastic Enclosures and ESD

All of the measures described above rely on the ability of a metal enclosure to dissipate the energy of an electrostatic discharge to ground. None of these methods can be used with all plastic enclosures. An all-plastic enclosure with no internal sheet metal, metal internal vaporization, or carbon particles—which limits the possible color range of the enclosure—is a case for a specialist. Contact an EMC test center early.

Air gap discharges at 15 kV are used for plastic enclosures. If several pulses are applied to the same spot with the same polarity, the charge accumulates and remains virtually there. If the polarity is reversed at another location, a large voltage difference can build up with respect to the first location. This can lead to a flashover right through the case, destroying the PCB components along the way.

Variable	IEC 61000-4-2 (ESD) Contact discharge	IEC 61000-4-4 (Burst)
Internal resistance	330 Ω	50 Ω
Maximum voltage	8 kV	2 kV
Energy	6 mJ	6 mJ
Waveform	(waveform with markers at 0 ns, 1 ns, 25 ns; U_{max})	(waveform with markers at 0 ns, 5 ns, 52.5 ns; 10%, 50%, 90%, U_{max})
Pulse rise time	< 1 ns	5 ns
Minimum pulses per pin	10 x positive, 10 x negative.	15 ms long packet with pulses, pauses 0.2 to 0.4 ms, repetition of the packet every 300 ms.

Table 24. Comparison of burst waveform with ESD waveform.

Burst Protection

Most readers will be familiar with the terms "burst" and "EFTs". Table 24 illustrates the form of interference compared to ESD. Electrical fast transients (EFTs) are common in industrial environments and are primarily caused by power interruptions to inductive loads. In EMC testing, bursts are capacitively coupled to all cables of the device under test (DUT) using a coupling clamp. If the data cable is shorter than 3 m, the test is omitted [R037].

It is interesting to note that the burst rise time is not orders of magnitude less than ESD, but only about a factor of five. The term "fast" is still justified. In addition, the energy in each pulse is equal to that of a standard ESD discharge. However, the maximum voltage applied is "only" 2 kV. The slower rise with the same energy at a lower maximum voltage means that the standard ESD protection provided by ESD avalanche diodes also protects against bursting, so no special additional measures are required [R050].

Surge Protection

The "surge" event, which simulates overvoltages caused by a lightning strike in the vicinity of the device, is described by the characteristics shown in Table 26. The potentially very high energy is remarkable. In the EMC test center, surges are capacitively coupled to all cables connected to the equipment that will be outdoors in the field. If an indoor cable is longer than 30 meters, the surge test is also applicable.

In addition to the 1.2/50 µs characteristic shown in Table 26, there is a characteristic with a rise time of 10 µs and a 50% time of 700 µs, which is used for symmetrically terminated connections such as telephone lines. For a common digital connection with a low impedance source and a high impedance load, the 1.2 µs/50 µs form should be used [R050].

Variable	Class 0	Class 1	Class 2	Class 3	Class 4	Class 5
Surroundings	Protected* cable	Partially protected cable	Good cable separation	Cables run parallel	Multi-core cable	Cable leaves building
Phase to neutral, or to VDC	-	-	0.5 kV (250 A)	1 kV (500 A)	2 kV (1000 A)	[1]
Phase or VDC to earth	-	0.5 kV (42 A)	1 kV (83 A)	2 kV (167 A)	4 kV (333 A)	[2]
Data line to earth	-	0.5 kV (12 A)	1 kV (24 A)	2 kV (48 A)	4 kV (95 A)	4 kV (95 A)
Data line to Data line	-	-	0.5 kV (12 A)	1 kV (24 A)	2 kV (48 A)	2 kV (48 A)

[1] Far away and well isolated from a cable leaving the building, [2] according to characteristics of the local mains network.

Table 25. Surge test levels [R053], [R052].

Variable	IEC 61000-4-5 (Surge)
Internal resistance	2 Ω between phase and neutral or VDC+/- (Differential Mode, DM). 12 Ω between phase or neutral and ground or between VDC and ground (Common-Mode, CM). 42 Ω between the signal line and ground or other signal line (> 30 m length)
V max.	4 kV
Energy	up to 90 J (!)
Waveform (idle)	So-called 1.2/50 µs shape
Waveform (short circuit)	As above with characteristic values 8 µs/20 µs
Pulse rise time	1.2 µs
Minimum pulses	5 positives and 5 negatives, pause of maximum 1 minute between each pulse

Table 26. Surge interference waveform

The type of surge protection depends entirely on the test level and whether the power and/or data lines need to be protected. Table 25 lists the test levels. In general, Class 3 tests are required for residential, commercial and light industrial applications. Class 4 tests are for heavy industrial applications only. But as with the ESD test levels, it is advisable to use Class 4 for any equipment where surge testing is performed.

Burst and Surge Protection Example

As mentioned earlier, if a cable is longer than 30 meters or exits the building, it should be surge tested. The example in Figure 62 (source: R050) shows that avalanche diodes alone do not provide sufficient protection because the current levels in a surge case would still be too high. However, sufficient protection can be achieved with additional elements:

- The TISP4240 thyristor protection device is capable of handling high surge currents, but only switches through above 250 V.

- Therefore, TVS diodes are still required, but they only clamp at +13.3 V and -7.5 V. However, this is acceptable because the ADM3485E transceiver allows a working common-mode voltage of -7 V to +12 V; therefore, the overvoltage load on its internal protection diodes is minimal.

- Finally, one more element is needed to allow the thyristor to carry the main current instead of the diode. This is the Transient Blocking Unit TBU-CA065. It is a 650 V safe current monitor IC that switches off at 200 mA in the event of an overcurrent.

The thyristor element has a response speed of 5 kV/μs. It is therefore suitable for surge pulses but does not block ESD pulses. Similarly, the TBU unit responds to overcurrent with a delay of 1 μs.

Overvoltages due to Long Cables

A long cable from the power supply to the PCB limits the inrush current due to cable inductance, which can be life-saving for tantalum capacitors on the PCB. However, the same cable inductance can become a serious hazard to the circuit if it quickly cuts off the source current. A practical case is shown in Figure 61: an "electronic speed controller" (ESC) is controlling a brushless DC motor. The ESC is connected to a power supply by a two-meter cable, and the distance between the ESC and the motor is one meter. If nothing is done, the ESC will fail sooner or later. This is due to overvoltages in the kilovolt range caused by the

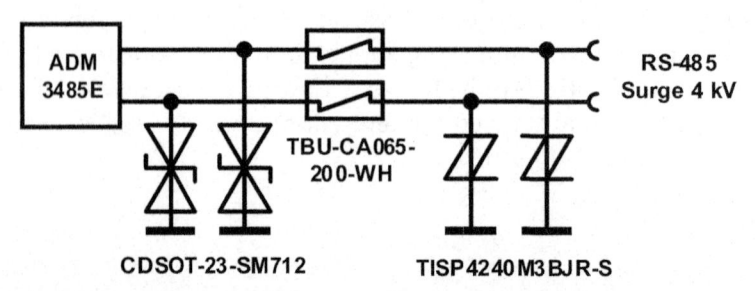

Figure 62. Protect an RS-485 connection from surge voltages.

fast switching of the ESC and the cable inductance. The cable inductance is small, 1 µH, but if the switching is fast enough, very high voltage spikes can occur. As a remedy, place a very large capacitor, e.g. 10 mF, in front of the ESC. Do not choose a tantalum type, however, this type of capacitor should be avoided at external power interfaces. The cable inductance may not limit the inrush current sufficiently, and tantalum capacitors are sensitive to sudden high currents.

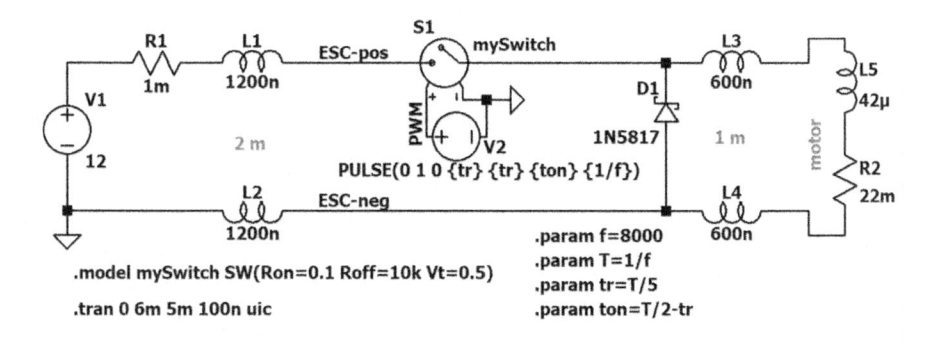

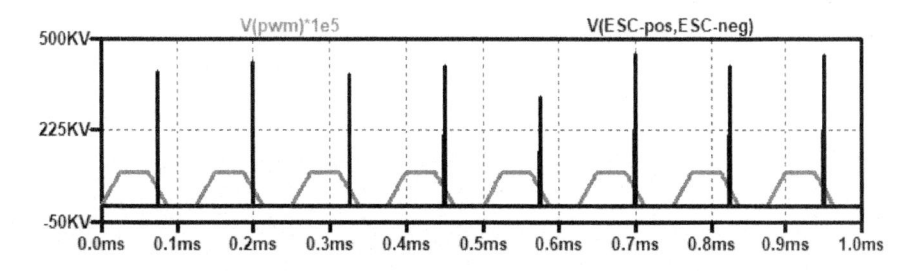

Figure 61. Circuit: Example of a case with overvoltages due to excessively long power supply lines, see text for explanations. Diagram: Overvoltage peaks across the switching transistor with each falling edge of the PWM switching signal, the switching signal shown is amplified 100,000 times for clarity.

Earth Potential Discrepancy and Differential Signals

Finally, when discussing safety, the problem of ground potential shifts should be mentioned. In a differential transmission pair, a ground potential shift between the two devices appears as a common mode noise (i.e., equally developed on both wires). Mathematically, this common mode noise is eliminated by the receiver building up the difference between the input voltages. In practice, a differential input stage cannot handle arbitrarily high common-mode voltages. Those greater than the input stage supply voltage are critical. Differences in ground potential (e.g. between buildings) can quickly exceed today's low supply voltages of 3 V or 5 V. Possible solutions are as follows:

- Differential receiver with high common mode voltage tolerance

- Galvanically isolated connection (e.g. via fiber optics)

These are remedies you can use. Far better is to eliminate the source of the potential ground shift in the first place. You may ask about this first. However, in my experience, people are reluctant to make major changes to a running system if it can be avoided.

Interference Protection Measures

We will now discuss the integrity of interfaces, that is, how they can be affected or how they can affect the environment without causing direct damage. The topic is Electromagnetic Interference (EMI), also known as Radio Frequency Interference (RFI). The basic EMI requirements are covered by the following international standards:

- IEC 61000-4-3: Specifying immunity to electromagnetic waves above 80 MHz.

- IEC 61000-4-6: Specifying immunity to fields and waves from 9 kHz to 80 MHz.

- IEC 61000-6-3: Specifying Household, commerce, light industry emission limits.

- IEC 61000-6-4: Specifying Industry emission limits.

Therefore, we are dealing with interference fields and radiation. All possible paths of interference for two connected devices are shown in Figure 63. As shown in Figure 63, stray currents will be discussed together with EMI in the following, since some EMI control measures favor stray currents. Note that direct interference into or out of the equipment is discussed in the next chapter.

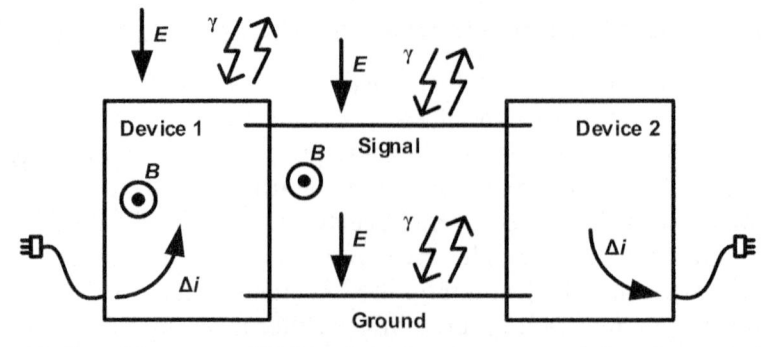

Figure 63. Possible cases of EMI in two devices connected with a signal connection (analog or digital). The strands of the signal wire are pulled apart, but in reality they are close together. The EMI interference mechanisms are an electric field E, a magnetic field B, electromagnetic radiation γ and galvanically coupled interference Δi.

Measures	G	E	B	↗	↙
Shorten cable length	O	☹	☺	☺	O
Choke	O	☺	☺	☺	☺
Three-terminal capacitors	O	☺	O	O	O
Common-mode choke, data line filter	O	☺	O	☺	☺
Twisting both conductors	O	O	☺	☺	☺
Differential signal transmission	☺	☺	O	O	😐
Cable shielding	☹	☺	O	☺	☺
Return conductor next to forward conductor	☺	☺	☺	O	O
Cable routing	O	😐	😐	😐	O

G: galvanic, E: electric field, B: magnetic field, ↗ transmitting, ↙ receiving.
O: negligible effect, 😐: effect depends on situation and design, ☹: increased interference level, ☺: reduced interference level.

Table 27. Overview of measures against EMC problems for connected devices.

Table 27 provides an overview of the measures commonly taken to improve the EMI situation or to suppress stray currents, including an indication of their effectiveness. Looking at Table 27, it is clear that there is no single solution to EMI. What is useful in one case may be useless or even counterproductive in another. Taken together with all the interference options shown in Figure 63, EMI appears to be a difficult subject. In fact, more than one type of interference and more than one location of interference are usually involved in a given case. Often, the proportions of each type of interference and the importance of each location cannot be measured separately or calculated theoretically. The coupling paths may not be obvious, may be counterintuitive, and may be distributed. In addition, the level of interference is highly dependent on the environment and the devices operating in the environment, such as motors, and the position and location of these devices. Finally, it is usually difficult to do anything about problematic interference without making significant circuit and layout changes. The bottom line is that EMI is a difficult subject.

In contrast the following can be said:

- Often, one type of interference is so dominant that it is sufficient to suppress it, and operation will resume without restrictions.

- Each type of interference has its own peculiarities that make it easier to estimate or recognize which type might be the dominant one. These characteristics are described below.

- There are effective preventive measures for each interference problem, even if they are not all compatible with each other, but this is usually not a problem, because in a concrete case not all interference types are equally pronounced.

- As a result, it is recommended that the maximum number of EMI measures be enabled in the prototype, if possible all of those listed in Table 27. Those that prove unnecessary in practical testing can be omitted from the final product.

The measures listed in Table 27 are now discussed in detail.

Choosing the Right Cable Length

Do not choose prototype cables that are unnecessarily long, but rather as short as possible. This recommendation is easily disregarded when components or modules are connected to the prototype. Often, an arbitrary length is used instead of the shortest possible. Also consider a direct connection instead of plug-cable-plug connections. In any case, have several cables of different lengths ready to perform connection length tests in the EMC test center.

The main purpose of shortening the cable length is to reduce radiation. Noise emission is at a maximum when as much power as possible is radiated. This is achieved with power matching, in the case of a transmitter, when the antenna impedance is equal to the source impedance.

Antennas that are very short compared to the wavelength have a high input impedance. There are conceivable high-impedance sources on printed circuit boards, such as crystals or certain resonant circuits. However, these are never operated at significant power because we do not have high supply voltages, and $P = V^2/R$ holds. Therefore, we get negligible spurious radiation from high-impedance sources, even if they happen to match an attached cable in terms of antenna matching.

The relevant spurious radiation occurs in conjunction with sources that are designed to have high power and, in addition, low internal source impedance.

For example, digital ICs belong in this category because they are designed to drive as high a current as possible into the capacitive load formed by the receive components to achieve as steep an edge as possible. Of course, all analog ICs and circuits with low internal resistance also belong to this category.

From the sources with low source resistance, we get a good spurious emission with structures that also have low impedance in the frequency band considered. Power matching is ideal, but relevant spurious radiation can occur even before that.

As a rule of thumb, significant spurious emission is likely when the trace length is about a quarter of the wavelength or more. The conductor does not have to have an elongated shape; only the length is considered in determining resonance, and the topology then determines directivity.

Also, a conductor grounded at both ends can be a good antenna. A characteristic of an antenna in resonance is that the current and voltage vary greatly along the antenna (i.e., an antenna is not a concentrated element, but a distributed one). This means that cables carrying strongly driven digital signals must be kept as short as possible. The limitation of cable length with respect to reflections in the absence of termination will be discussed in the next chapter.

Let us look at the effects of shortening a cable with respect to the types of interference not yet discussed with this focus:

- Shortening the cable length helps against magnetic coupling by reducing the loop area.

- Shortening leads to higher coupling of electric fields because the cable capacitance decreases and the cable impedance increases.

- Shortening may increase stray currents, but not if the cable has comparatively low resistance to other paths.

- Regarding reception, it is not necessary to get as much antenna power as possible, because we have the possibility to amplify the received signal. Therefore, we do not need the usual low impedance associated with power matching for a receive input. Rather, we want the input impedance to be as high as possible in order to get the highest possible signal voltage. This allows us to receive signals without antenna resonance. For interference reception, the length and dimensions of the structure acting as an antenna are therefore of secondary importance. Shortening the cable will not do much to reduce interference. Reduce the impedance level if you are experiencing interference problems.

Chokes in the Signal Path?

The previous chapter discussed the use of chokes and ferrite beads in the supply path. A choke can also be used to attenuate harmonics on a digital signal. For space and cost reasons, you will not want to attenuate entire buses in this way. But why not for a clock signal, so that its harmonics are not radiated through the clock network?

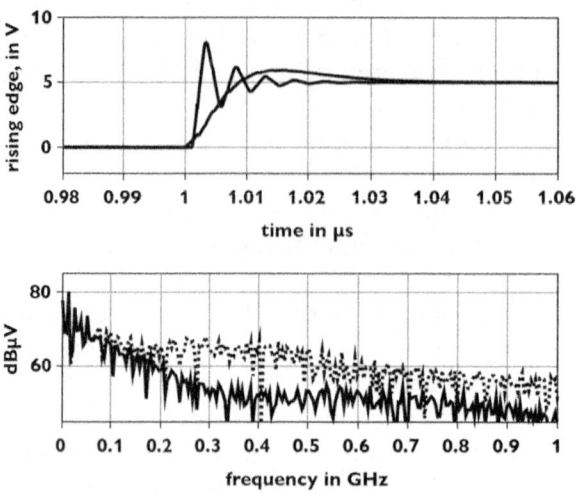

Figure 64. Example of the effect of a choke based on a 20 cm connection between an AC04 inverter (2 ns rise time) driven at 1 MHz and an AC00 NAND gate input. The curve without the choke shows strong overshoots due to the long line, resulting in a high spurious spectrum level. With the BLM18RK221SN1 (220 Ω @ 100 MHz) choke inserted, the transient and the spurious spectrum in dBμV are reduced by up to 25 dB in places from about 200 MHz [R142].

The choice of specific choke in the signal path is a trade-off between harmonic attenuation and acceptable rise time delay. Figure 64 shows an example of how a choke effectively suppresses the high-frequency noise spectrum in the signal path between the two AC family logic devices. Such a measure inherently increases the rise time of the signal. What is acceptable depends on the application.

Using Three-Terminal Capacitors

Another element used to filter out high-frequency noise is the three-terminal capacitor, also known as an EMI C filter, feedthrough capacitor, or T filter. To effectively eliminate high-frequency noise with a low-pass filter, the capacitor used must have as high a self-resonant frequency as possible. By designing capacitors in a specific way (see Figure 65), the feedthrough inductance can be minimized and the resonant frequency can be increased.

Compared with a conventional 0805 size X7R capacitor with 1 nF capacitance and a resonant frequency in the 200 MHz range, a three-terminal capacitor of the same size and capacitance has a resonant frequency of 500 MHz. Smaller capacitors can increase the resonant frequency by a factor of five or more.

Three-terminal capacitors are not ESD safe, so they should not be the first element after a connector.

Some engineers have had bad experiences with T-filters: if they break, the result is a short to ground and the device stops working. Dirt that accumulates around T filters can cause a partial short because the ground pins of this type of capacitor are close to the signal pins.

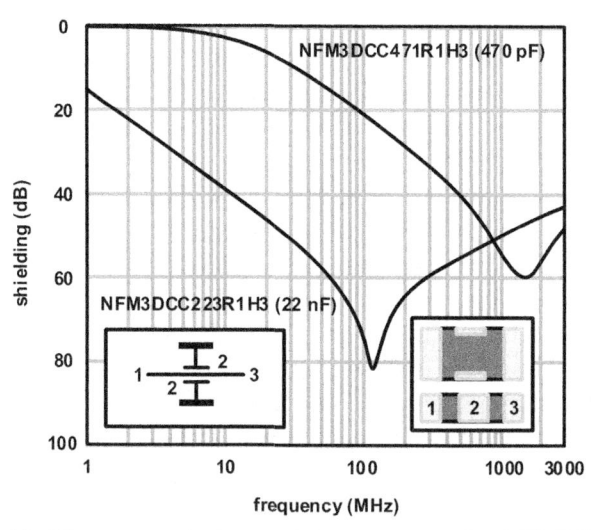

Figure 65. Three-terminal capacitors, such as the NFM3DCC223R1H3 type, have a capacitance of $22 \cdot 10^3$ pF = 22 nF (source: Murata). Insertion Loss: Amplitude response when the source and load have a resistance of 50 Ω.

Using Common-Mode Chokes

Common mode chokes (CMCs), also known as data line filters, are the first choice for suppressing vagabond currents. In its simplest form, a common-mode choke consists of a hinged ferrite that encloses the forward and return lines and is effective from about 100 MHz. In more complicated forms, counter-rotating windings are applied to a core. These are designed to be effective from about 100 kHz. CMCs are also available in SMD packages.

A common mode choke will resonate at a certain frequency. If a significant portion of the current drawn is at this frequency, the situation

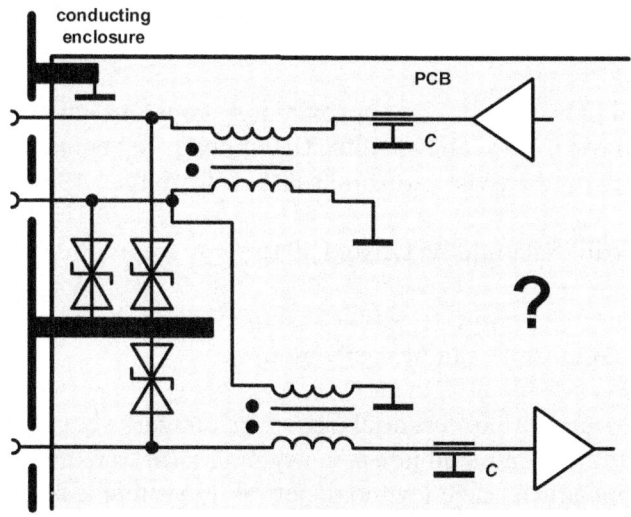

Figure 66. Single-ended connection with CMCs.

can be worse than without a CMC. Above the resonant frequency, the common mode choke is capacitive due to the leakage capacitance between the windings; (i.e., it is ineffective against common mode interference). It is therefore advisable to make the choke bridgeable, especially for the first prototype.

Figure 66 shows an application for the use of CMCs. There is a question mark, however, which indicates a problem. With the fast edges of today's digital signals, CMCs alone are no longer sufficient to suppress parasitic currents. The higher the frequency, the better the path through the CMC's winding capacitors. For high-speed interfaces, the solution shown in Figure 66 is therefore obsolete. It is worth noting that sticking with an apparently old and slow interface IC type will not help; "old" ICs are still being manufactured in state-of-the-art fabs and have much steeper edges today. The best solution is to use differential data transfer with CMC, as will be shown briefly.

Cable Twisting and Differential Transmission

By twisting a forward conductor and the corresponding return conductor, you cause a counter-induced voltage in adjacent twists so that the total induced voltage cancels out. Twisted-pair cables are an extremely effective method against inductive coupling and cause less radiation and interference pickup by reception as long as the wavelength is large compared to the winding length [R056].

All cables, including DC power cables, should be twisted if possible, since it is the flow of current, not the varying voltage, that is important, and supply currents can also have high-frequency components.

A few notes on twisting:

- Crosstalk can be reduced by winding adjacent twisted pairs in opposite directions.

- Winding adjacent twisted pairs with different twist ratios, such as 2:1 or 3:2, can also reduce crosstalk.

- A twisted pair cable has an approximately constant impedance of 100 Ω up to 10 MHz, above which the impedance begins to vary depending on the twist ratio and twist uniformity.

- Twisted cables should be twisted all the way to the ends.

- There are "twist-and-flat" cables that are flat-ribbon at the ends so that IDC connectors can be easily mounted.

Twisted-pair cable with differential transmission works best for signals and to suppress radiated noise. In asymmetrical transmission, the potential on one line is varied while the other line remains at ground potential, which is usually the same as earth potential, except in the

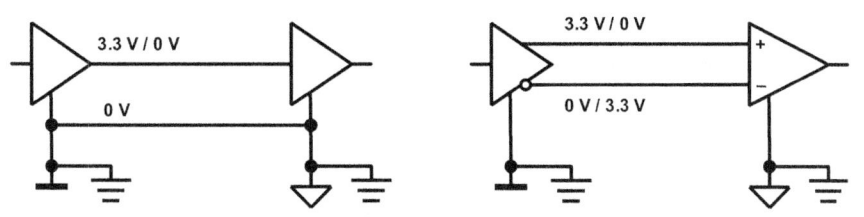

Figure 67. Asymmetric transmission (left), differential transmission (right).

case of double-insulated devices (see Figure 67). In differential transmission, also known as "balanced" transmission, the potential on both lines is changed in a push-pull fashion (i.e., relative to ground or earth potential); the lines are alternately at the HIGH or LOW level for a signal. This does not necessarily have to be a positive/negative voltage, both logic level voltages can be positive.

Differential transmission over twisted-pair cable is the absolute standard today and is the only way to achieve today's high data rates without EMI interference. Differential transmission standards are discussed at the beginning of this chapter.

Cable Shielding

Now we come to another important method of avoiding EMI problems: shielding. Shielding is now mandatory for enclosures, otherwise you will have problems with radiation directly from the device. This will be discussed in the next chapter.

Here we consider cable shielding (see Figure 68). If the cable shield is connected to the chassis at both ends, electromagnetic radiation and electrical field interference can be completely suppressed. However, when connected at both ends, a low-impedance ground connection is created (i.e., a ground loop with all other ground connections), and you may get problems with inductive coupling and stray currents. If the cable shield is connected to the chassis on only one side, it acts like an antenna and receives electromagnetic waves. The shield then still serves to attenuate interference from the electrical field. Therefore,

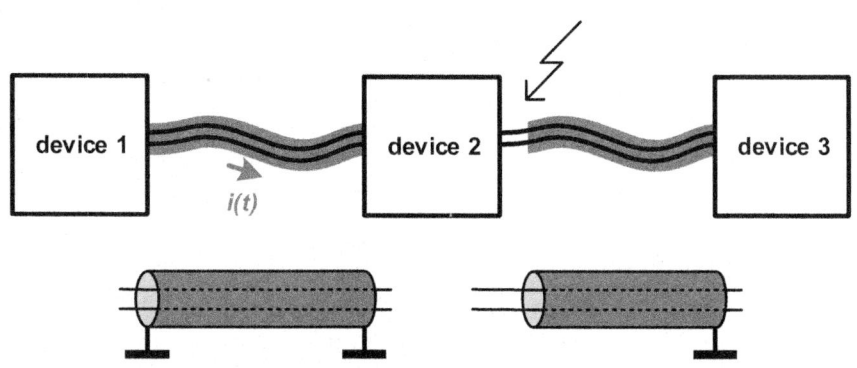

Figure 68. Cable shielding: Double sided or single sided?

Figure 69. Attaching the cable shield on one side: which side?

there is no clear instruction on how to use shielding in general; you have to find out which interferences cause the most problems in your specific case. You should provide a chassis connection on both ends with a proper cable, but you should also provide one that is grounded on one end only. In the EMC test center, you can then perform experiments with both cables.

Cable Strands and Cable Routing

Providing a ground wire for each unbalanced signal wire next to it would have advantages in principle—see Table 28—but is usually not feasible because the cable and connectors become too wide and too expensive. For flat ribbon cables, it is recommended that at least one ground wire be placed next to ten signal wires, with the ground wire in the middle (see Figure 70). Each additional ground wire has an effect on electric and magnetic fields and radiation as shown in Table 28. Additional ground wires also prevent coupling between the signal lines, which is often the main problem. Therefore, the following recommen-

EMI Type	Judgment	Justification
Galvanic coupling	☺	When each signal has a return wire, there are many ground wires that provide a low impedance path, reducing stray current problems.
Electric field	☺	The capacitance of the line pair is increased, reducing stray electric fields and electric field coupling.
Magnetic field	☺	The loop area is reduced, magnetic fields are reduced, and magnetic field coupling is reduced.
Electromagnetic radiation	O	At high frequencies, ground wires can also act as antennas. No improvement, but no detrimental effect to expect either.

Table 28. How a ground wire next to a signal wire affects EMI.

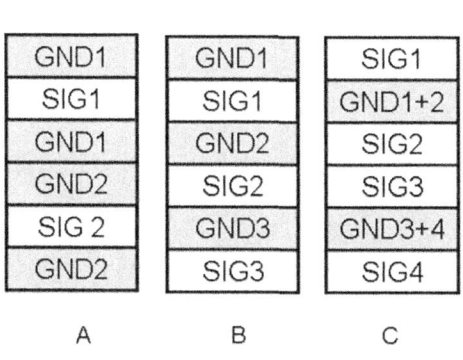

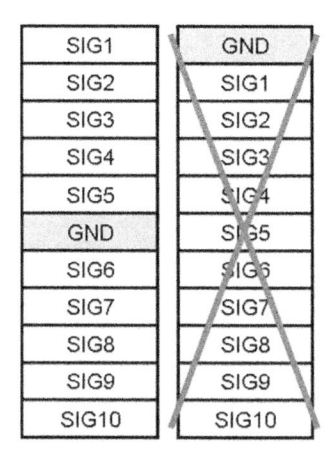

GND1	GND1	SIG1
SIG1	SIG1	GND1+2
GND1	GND2	SIG2
GND2	SIG2	SIG3
SIG 2	GND3	GND3+4
GND2	SIG3	SIG4
A	B	C

SIG1	GND
SIG2	SIG1
SIG3	SIG2
SIG4	SIG3
SIG5	SIG4
GND	SIG5
SIG6	SIG6
SIG7	SIG7
SIG8	SIG8
SIG9	SIG9
SIG10	SIG10

Figure 70. Ground wires on the ribbon cable.

dations are made regarding the use of ground wires in flat ribbon cables (see Figure 70):

- The best arrangement is variant A.

- The second best arrangement is variant B. If all grounds GND1...n are connected, it is practically as good as variant A.

- Variant C is tolerable.

Clock signals and high-speed signals should always have a ground wire on both sides.

In addition to the data lines, power is often carried on a flat ribbon cable. This brings up the issue of what is allowed on each strand of the cable. The main concerns here are crosstalk and interference between the wires.

Basically, there are groups of signal types that are compatible with each other and therefore can be routed on strands of a cable without any problems. However, this also means that routing signals from different groups can cause problems. The groups are as follows:

- AC power supply and ground, chassis ground, power audio signals and ground.

- DC power supply and its ground, voltage references, sensitive audio signals and their ground.

- Digital signals and their grounds.

- High frequency power signals and their ground.

- Low frequency signals and their ground.

- Antenna feed line: high power and very sensitive intermittent.

Cables from different groups should not even be routed in the same cable harness. However, if parallel routing is required, minimum distances are recommended:

- 2.5% of the length of the parallel section between digital signals and power lines carrying high currents, either AC or DC.

- 25% of the length of the parallel section between analog cables carrying millivolt signals and power lines, whether AC or DC.

If you cannot avoid mixing signals from different groups, it is recommended to do so, as shown in Figure 71:

- Provide a ground wire for each signal wire.

- Connect additional ground wires between signals from different groups.

Flat ribbon cables installed on top of each other should be kept apart with plastic spacers or a plastic mat that is at least as thick and as wide as the cable.

Route the cables inside a case based on the following EMI criteria:

- Cables should be routed along the inside of the chassis, bypassing all chassis openings as far as possible—radiation occurs through these openings.

- Do not route cables near transformers, chokes, motors, or coils.

- Cables should not be routed over microcontrollers, oscillators, or other intensive signal sources. This can easily happen when a ribbon cable is folded into the enclosure during assembly.

- If possible, all cables should be routed radially away from the board, and no cables should loop over the board.

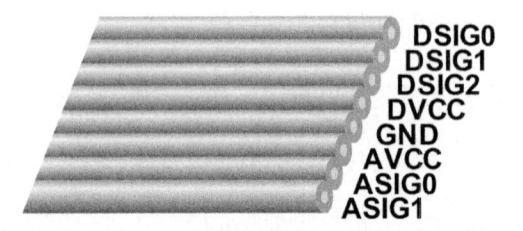

Figure 71. Digital and analog signals on the same flat ribbon cable.

Interface Example Designs

We will now consider and discuss different implementations of the above for asymmetrical and symmetrical interfaces. Figure 72 shows an example of interference suppression for an asymmetrical single-ended interface. In the single-ended interface, data is transmitted as voltages with respect to ground. The T-filter, in conjunction with the choke, suppresses high-frequency interference and noise reception. This interaction and its specific elements can only be fine-tuned experimentally. It should be noted that it is recommended to use a choke in the ground line as well. This is because the local ground can be very noisy with digital signals [RO98]. Check that there is no saturation.

The question mark in Figure 72 is placed because, in this example, there is a high risk that some of the return current will not take the intended path due to alternative signal ground connections or, for example, the main ground (see Figure 73).

In addition to the problem of uncontrolled return paths, there are also problematic ground loops with multiple ground connections (see Figure 74). Large ground loops of this type are problematic with respect to induced interference. Especially for an analog circuit, unacceptably high noise can be generated on the ground.

Better use symmetrical signals instead. A generic layout for the interface with symmetrical signals in the form of the maximum variant is shown in Figure 75:

- Separate area for filtering with its own ground plane, which should be as close as possible to the front panel and chassis due to electrostatic discharge.

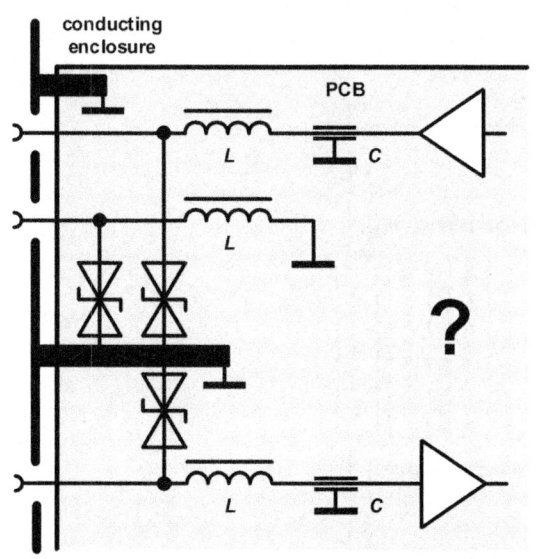

Figure 72. Choke and T-filter in combination for a single-ended interface. A choke is also recommended on the ground line. The question mark is there because adding more ground connections can cause stray currents.

136 Robust Interfaces

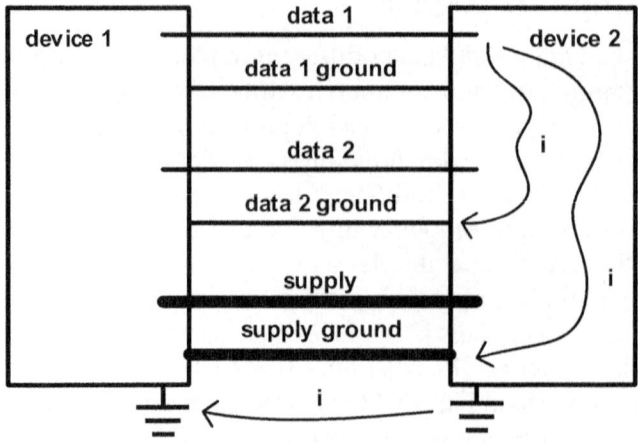

Figure 73. Vagrant return paths for fast single-ended connections.

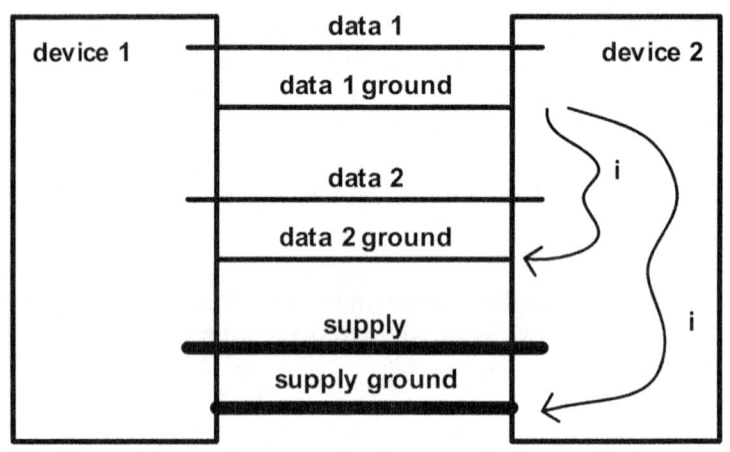

Figure 74. Ground loops due to multiple ground connections.

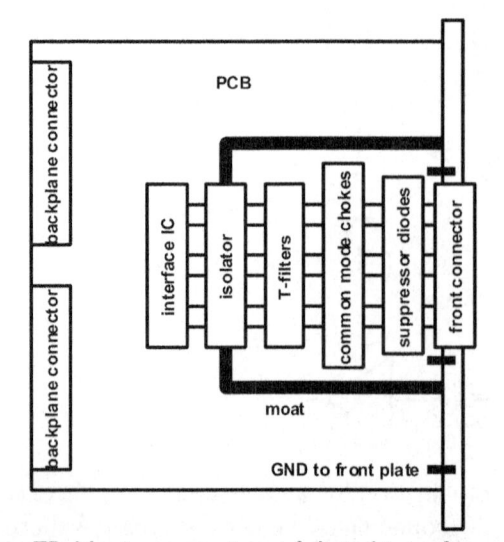

Figure 75. Maximum variety of digital interfaces.

- Avalanche diodes or ESD protection module as close to the connector as possible.

- Common mode choke.

- T-filter. Placement before or after the CMCs is possible if protection diodes are used.

- Galvanic isolator: This is the most expensive component of this maximum variant, especially if it is not only receiving and therefore requires a bidirectional isolator. If you reduce the maximum variant by the isolator, you can approximate the isolation with an additional voltage regulator or by adding ferrite beads to the power supply of the interface IC.

- Differential driver/receiver IC.

Figure 76 shows a USB front end as an example of a fast digital interface. Except for the galvanic isolation, this is the maximum variant of Figure 75. Figure 77 shows the generic layout for a minimum variant:

- Use a bridge in the ground plane instead of the insulator.

- Protection diodes can be omitted if the interface IC has integrated protection diodes according to HBM. The energy reached by the IC is already sufficiently reduced by the common mode choke and the distance (i.e. the inductance of the traces).

- There is no ground plane under the data lines between the common mode choke and the connector. This prevents energy from being coupled into the ground plane during the ESD pulse.

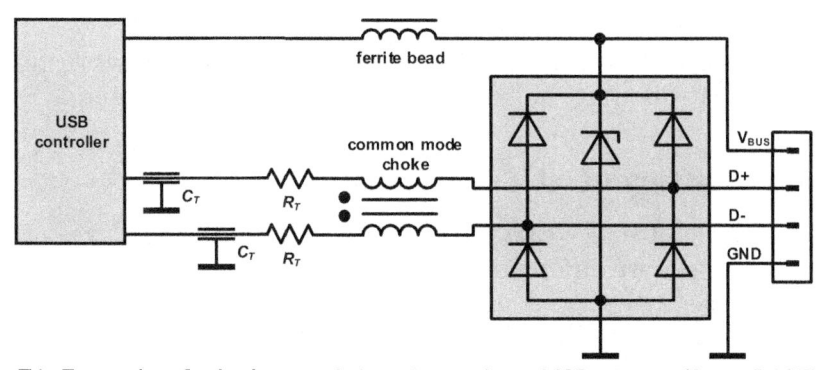

Figure 76. Example of a high-speed digital interface: USB circuit (from R097). The terminating resistors R_T and the T-capacitors C_T form a low-pass filter.

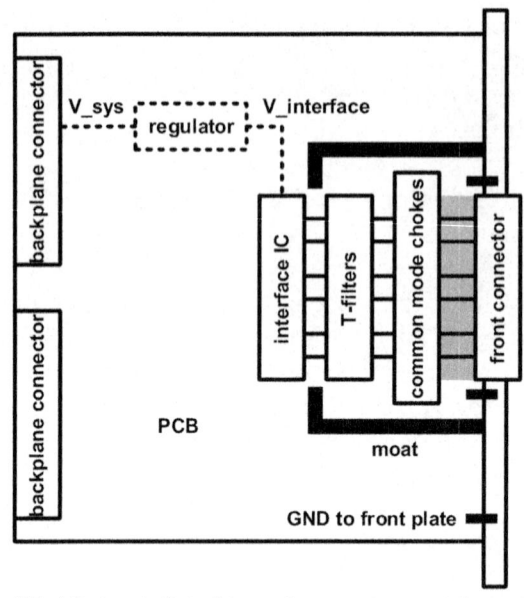

Figure 77. Minimal digital interface variant with optional dedicated power supply for the interface IC.

- The common mode choke may be omitted, but it is recommended to keep it in place as it attenuates the ESD pulse and does not change the waveform.

- Important: If T-filters are used, place them after the common mode choke; this will protect the T-filters from the ESD pulses. The T-filter capacitors are not ESD proof.

- Connection to the front panel is provided by the connector itself, using not only a pin, but also proper conductive mounting brackets connected to ground.

- Differential driver/receiver IC.

Optionally power the interface IC from a separate voltage regulator to eliminate interface noise from the IC to the PCB power supply.

Analog Differential Transmission?

What about analog signal transmission? Analog, ground-free AC voltage signals are differential signals by nature. In practice, however, it is necessary to ensure that AC signals are truly ground-free. In a microphone, a differential signal is naturally created by the vibration of the diaphragm, which produces both positive and negative voltages with respect to ground. However, since the microphone often has a metal housing, more expensive microphones have a signal transformer built in to prevent a ground loop across the microphone mount, see e.g., the

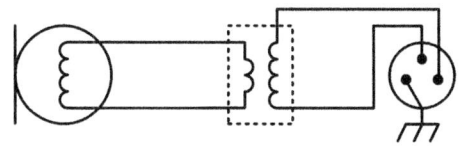

Figure 78. Internal schematic of the Shure Model BETA 58A microphone.

Figure 78. Such microphones can be identified by the XLR connector, which has a ground, positive, and negative pin. This is a true differential transmission of an analog signal. Before attempting this, however, it is important to determine whether a simple low-pass filter at the input would not be sufficient. If not, a conversion to an AC voltage using a voltage-controlled oscillator (VCO), e.g. the MC14046B, seems relatively simple. However, the receiver side is much too complex. Instead, the simplest overall method is to generate a pulse-width modulated signal (PWM signal) in some kind of A/D conversion, e.g. with the TL494 device, with an accuracy of about 5%. On the receiver side, the DC signal can be recovered relatively easily with a low-pass filter. Differential transmission may no longer be necessary. If it is still required, it can be implemented with a differential digital driver so that the PWM is transmitted symmetrically. A high calibration effort has to be considered. For a sensor, the question arises whether it is not possible to convert the measured quantity directly into digital values at the peripheral device. If the peripherals are powered anyway, check if a sensor with a digital interface is available instead of an analog sensor.

With an analog signal, the desired frequency band can be filtered out with a specific bandpass filter. The filter design is unproblematic up to 10 MHz [R185, R199 and R200]. Protective diodes, an additional ground plane, an isolator or bridge, and a good connection of the additional ground plane to the front panel/chassis should be provided, see Figure 79.

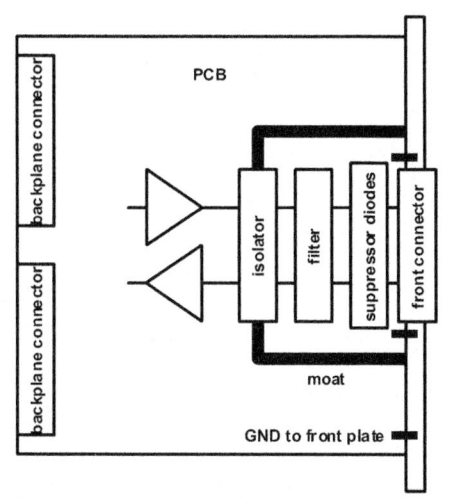

Figure 79. Analog, low-frequency interface.

Some Advice on EMC Testing

At the end of this chapter, I would like to offer some recommendations for EMC testing:

- Rather than waiting for the first prototype to be assembled, some EMC test centers allow review at the schematic level or with PCB layer printouts. Their suggestions are usually worth the money. EMC test centers are accustomed to signing a non-disclosure agreement, but usually non-disclosure is part of the laboratory's professionalism.

- Perform the EMC tests that you can do yourself, which are listed in the chapter on Testing and Verification, as early as possible.

- Have the first prototype EMC tested. Often there is a need for improvement. These can usually be incorporated easily at this early stage. Testers can also provide valuable advice on how to improve EMC beyond what is necessary.

- Allow for maximum EMC measures in the first prototype. This allows you to use the valuable time with the EMC expert not only to pass the test, but also to optimize the design. A typical example of this is the ability to put a choke in a path. If you do not provide for this, it may not be done quickly or at all at the test center.

- For initial EMC testing, go to the largest EMC center in your area. They have the largest knowledge base, and there is a good chance that the tester will be an expert in the field of your device.

- You do not have to go to the large, usually expensive EMC center later, later on. If you just want to repeat passed tests with a new prototype series, you can go to a smaller EMC test provider that can perform the tests. These are often less expensive.

4 Signals on Printed Circuit Boards

A signal on a printed circuit board (PCB) can be disturbed or interfered with from the outside; this is the subject of signal integrity, shielding, and board ground loops. A signal on the PCB can interfere with another signal on the same PCB; this is the subject of stability, crosstalk, and mixed-signal designs. Finally, a signal on the board can interfere with itself, and then we talk about reflections, intersymbol interference, and overshoot. This chapter is organized according to the list above.

Enclosure Shielding: Ideal and Real Faraday Cage

Shielding the enclosure is the most effective way to block external interference and prevent unwanted emissions from the PCB to the environment. Electric fields and electromagnetic radiation can be shielded relatively well by a thin aluminum foil. Attenuation is caused by reflection and absorption due to the skin effect (see Figure 80). Depending on how close we are to the transmitter, we are in the near or far field. In the far field, the ratio E_t/B_t of the tangential electric field to the tangential magnetic field is constant at any location. In the near field, the ratio depends on location and time. In one limit case, we have a practically pure electric wave; in the other limit case, we have a practically pure magnetic wave. The impedance of a pure electric wave is high: high voltage/low current. As the impedance of the screen becomes smaller with lower frequency, there is more and more reflection—as with strongly differing refractive indices in optics.

A purely magnetic wave has a low impedance that matches the low impedance of the screen, so the reflection decreases in strength as the frequency decreases. At low frequencies, the electric and magnetic

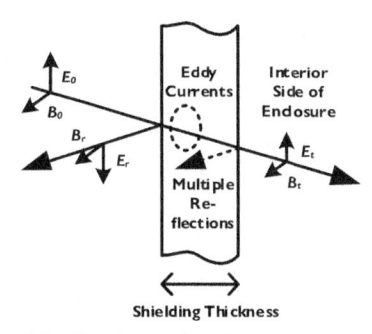

Figure 80. Shielding mechanisms.

fields are not coupled, and the attenuation of the fields is different; we are in the so-called "near field" of the source.

Shielding is considered "good" when the attenuation is 60 dB or more. The more conductive the shield and the lower the permeability of the shield, the more effective the reflection attenuation. Transmission attenuation (i.e., absorption by eddy currents) increases exponentially with the thickness of the layer [RO96].

A Faraday cage made of normal, non-magnetic metal is not field-free, but allows static and low-frequency magnetic fields to pass through. This is often misunderstood because commercially available Faraday cages also attenuate low-frequency and static magnetic fields, but they do so by using a high-permeability layer of so-called "mu-metal". However, mu-metal is expensive and, above all, pressure-sensitive, so it can only be processed with great care. In practice, it is not possible to provide mass-produced devices with mu-metal shielding.

When shielding enclosures, it is important to note that the shielding effect is largely a function of the openings present. As the frequency increases, the wavelength of a signal decreases, which means that even small openings become large "gates" for high-frequency signals. For wavelengths λ less than or equal to twice the largest aperture length d (Figure 81, $\lambda \leq 2 \times d$), there is no attenuation. For wavelengths greater than twice the largest aperture length, the attenuation increases linearly at a rate of 20 dB per decade. Comparing the attenuation of the Faraday cage and the attenuation of the apertures shows that the total attenuation is essentially determined by the attenuation of the apertures. The largest dimension of the aperture is decisive for the calculation. For example, for frequencies up to 1 GHz and a minimum shielding of 20 dB, the largest opening must not be larger than 10 mm, as shown in Figure 81.

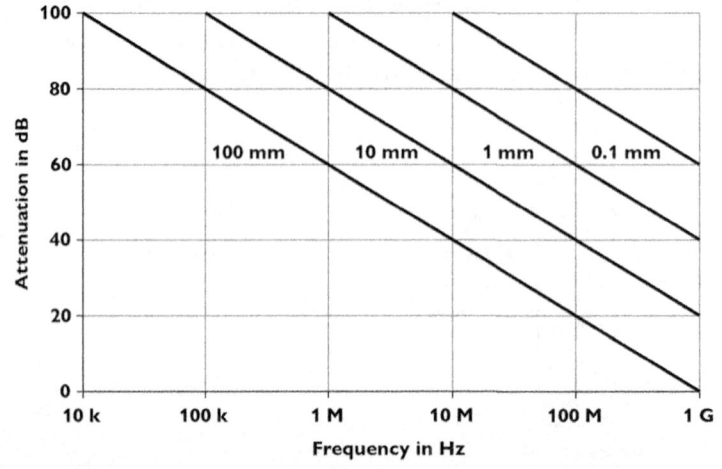

Figure 81. Enclosure openings and resulting attenuation reduction.

Ventilation holes can be covered with a perforated plate without major problems. If the hole spacing is less than $\lambda/2$, the overall shielding is worse than the shielding of a single hole by a factor of the square root of the number of holes. This results in the following: For example, if there are two holes, the amplitude of the incoming radiation is twice that of a single hole. The intensity scales as the square of the amplitude. The incoming radiation is diffracted at the perforated plate as at an optical grating and superimposed into a total intensity. Therefore, the total intensity is proportional to the square of the number of holes. An example: One hundred 4 mm holes shield worse than the single 4 mm hole by a factor of $\sqrt{100} = 10 = 20$ dB.

Two identical apertures separated by more than $\lambda/2$ do not cause a significant increase in shielding loss [R057]. Diffraction still occurs, but there is no significant superposition. We measured the same radiation amplitude near one aperture as at the location of the other aperture.

How to Manage Openings

Here are some ideas for dealing with emissions issues:

- Displays with very large apertures can be covered with transparent conductive film.

- Sensitive circuit parts or circuit parts with high interference potential should be placed away from the openings.

- Avoid gaps by using overlaps and spring contacts; electrical contact is helpful. Clarify surface treatment and possible oxidation [R098].

Ground Plane Advantages and Problems

In the chapter on robust interfaces, we discussed the EMI advantages of routing the return conductor of a signal directly next to the forward conductor. In fact, the standard for high-speed signals is to twist the forward and return conductors to eliminate inductive coupling.

There is no way to twist the PCB. What you can do is place a ground plane as close as possible to the forward current. According to the 2H rule from the supply design chapter, the closer the forward current to the ground plane, the more concentrated the return current will be under the forward current. However, this is easier said than done. First, the spacing of the layers usually cannot be chosen differently from a given standard setup without additional cost. Second, if a power plane is used, the forward current is split into a portion that flows on that plane and a portion that flows in the signal layers. Figure 82 on the left shows a typical layer stackup for a four-layer board. As can be seen, it

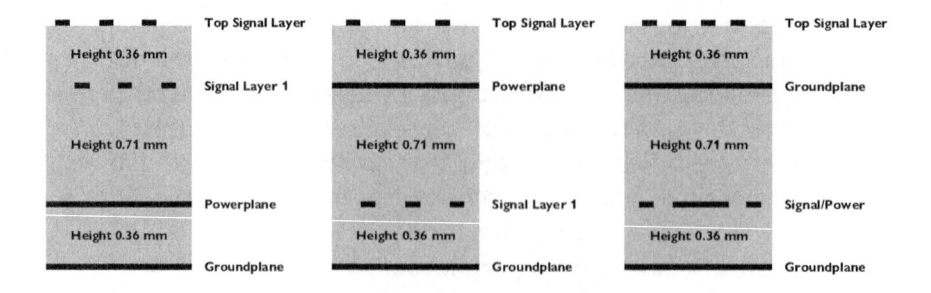

Figure 82. Examples of board layouts. The example on the left has a built-in low-inductance power-ground plane capacitor, which is preferred for supply decoupling, but the forward currents in the top and inner signal layers are far from the reference planes. The latter is partially improved in the middle example, but we have lost the low-inductance plane capacitor. The example on the right regains the plane capacitor by flooding the inner signal layer with power, which also allows for a second ground plane directly below the top layer.

is not possible for all signal layers and the power plane to have a minimum distance to the ground plane. This can cause a device to fail the EMC test even though it has a ground plane.

The example in the middle has an inner signal layer with a return path directly below it at minimum height. However, the forward currents in the power plane are now far from the return currents.

A solution is shown in Figure 82 on the right. The inner signal layer is combined with the power plane to form a common power/signal plane. This works well if the inner signal layer has only a few traces. You may need to move some traces to the top layer. In the chapter on power supply design, an example is given where a device passed the EMC test using only flooded signal layers. However, both flooding and higher trace density can lead to increased crosstalk if not done correctly, which is discussed in the next section.

Today, the ground plane is so standard that you may not even think about its disadvantages. However, this should not hide the fact that the ground plane also has disadvantages compared to an explicit return conductor placed right next to the forward conductor:

- A ground plane is a giant galvanic coupling conductor, and components in close proximity to each other can be expected to experience galvanically coupled noise through the ground. In fact, this is one of the main problems with very fast circuits with their high current pulses.

- With the ground plane, you have created a huge area where an external magnetic field can couple. Interference is not shorted through the ground plane any more than eddy currents are shorted through metal plates! External magnetic fields induce voltages and currents that can make the ground noisy.

Figure 83. Coupling capacitances and inductances.

How to Avoid Crosstalk

We now turn our attention to the stability of signal connections on the PCB and begin to discuss crosstalk between digital signals. Crosstalk occurs when signal components on one trace cross over into another trace via the capacitance to another trace or via the coupling inductance to the other trace (see Figure 83). The effect is more pronounced the longer a transient condition exists on the line (i.e., the effect is most pronounced for signals with short rise times on long lines).

The most effective measure against excessive crosstalk is to increase the distance between lines that must be routed in parallel. The "3W rule of thumb" can be used as a guide. Two uncorrelated traces should be at least three times the width of the widest trace apart, center to center. However, you lose a lot of space with the 3W rule (see Figure 84). Note that the distance must be maintained not only to the next trace on the same layer, but also to the next trace above or below.

To justify this rule, we can look at the odd-mode impedance of one trace as a function of the distance to the second trace (see Figure 85). It can be seen that when the lines are 0.2 mm above the ground plane, the odd-mode impedance hardly changes beyond the 3W distance and then becomes the same as if the second line were located at infinity. This means that mutual interference has become negligible. An appropriate calculation in R121 shows a coupling of 5% for microstrip and 2% for stripline with the same setup. However, Figure 85 also shows

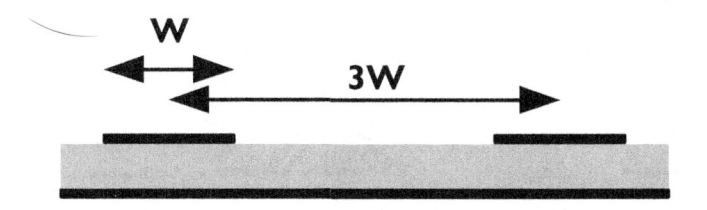

Figure 84. 3W rule against excessive crosstalk. The scaled drawing shows two traces on an outer layer with 35 μm copper overlay, a 200 μm prepreg, and a ground plane on the first inner layer. The other layers are not shown.

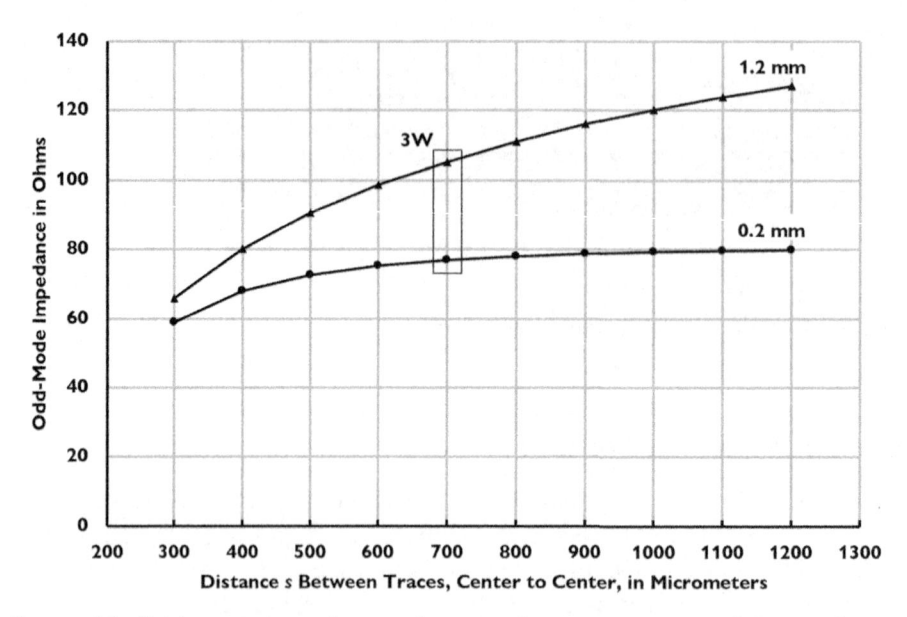

Figure 85. Odd-mode impedance of a pair of microstrips, one 1.2 mm above ground plane, the other 0.2 mm. The traces are 0.2 mm wide and based on a 17 μm copper layer. A dielectric constant of 4 has been assumed for the prepreg and FR4 materials. The curves are based on formulas from [R188].

that the 3W rule does not apply well to a signal height of 1.2 mm above the ground plane. The corresponding calculation in R121 gives a coupling of about 30% for microstrip at the 3W distance, which is of course too high. Since crosstalk is an AC problem, the ground plane can also be replaced by a power plane, since all power supplies are connected to ground with low impedance decoupling capacitors. At

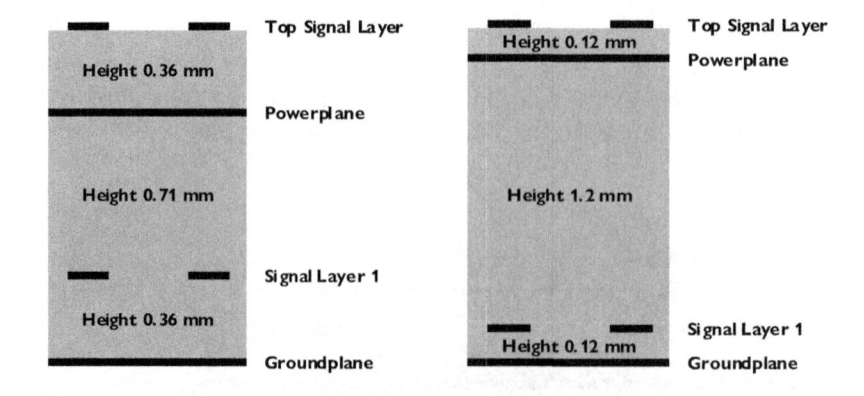

Figure 86. 4-layer PCB with standard structure on the left, with improved layer structure for reduced crosstalk on the right, sometimes offered as "standard impedance controlled". Outer layers with 35 μm copper overlay, inner layers with 18 μm copper overlay, and total thickness with lacquer in both cases 1.55 mm.

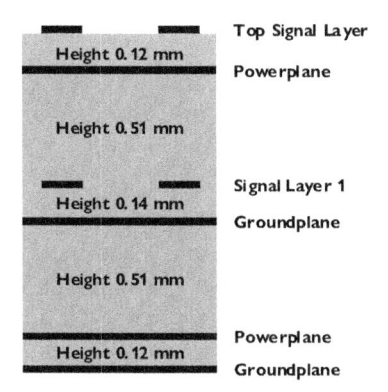

Professional Electronic Design Best Practices 147

Height 0.12 mm — Top Signal Layer / Powerplane
Height 0.51 mm
Height 0.14 mm — Signal Layer 1 / Groundplane
Height 0.51 mm
Height 0.12 mm — Powerplane / Groundplane

Figure 87. 6-layer printed circuit board with matched layer structure for low crosstalk, including effective ground supply area capacitor.

frequencies above 200 MHz, the capacitance of a decoupling capacitor acts directly as a low impedance connection, provided it has been designed effectively (see the section on DC supply design). Now, a good DC supply decoupling design for a typical 4-layer board comes into conflict with the requirement for the smallest possible distance between signals and power planes. The DC supply decoupling design also requires the power and ground planes to be adjacent. However, with the heights given in a standard 4-layer stackup (see Figure 86), only two of the three heights are reasonably small. The usual solution is to place signals in the power plane at non-critical locations. With the 4-layer board, you effectively have about 2.5 layers available for the signals.

With a 6-layer board, it is easier to meet both the crosstalk and DC supply decoupling requirements because there is the option of an additional ground plane power plane capacitor (see Figure 87).

Crosstalk can occur at the near end or the far end (see Figure 88). Near-end crosstalk is at its maximum at the so-called critical trace length and remains at that level for even longer traces. It can only be reduced by increasing the trace spacing. As a rule of thumb, the critical trace length, not to be confused with the trace length limit for termination, is about 12.7 cm for a rise time of 1 ns and a dielectric constant of the PCB material of $\varepsilon r = 4$. Far-end crosstalk increases much more slowly with length than near-end crosstalk, about 1/10. We usually do not have problems with far-end crosstalk on the PCB; the trace lengths are too short.

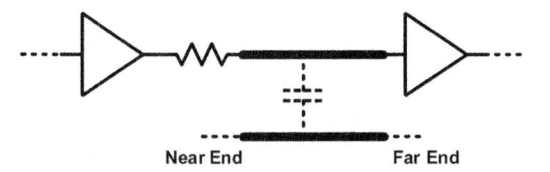

Figure 88. Near end and far end crosstalk.

Flooding Signal Planes

Flooding signal planes with ground or supply potential is clearly advantageous in terms of decoupling the power supply and preventing magnetic coupling. However, flooding can increase crosstalk. The greatest distance between two points on the flooded plane where a connection to ground or a power supply plane is made, either by a via or by a capacitor to ground, determines the frequency at which that plane can resonate. Since both ends are connected to the same potential, it is a half-wave resonance and multiples of it. Let us assume that a plane flooded with ground is connected to the ground plane at a maximum distance of 2 cm, with a via at each end. This results in a wavelength of 4 cm. With a propagation frequency of about half the speed of light, more on this assumption below, we get a resonance frequency of 3.75 GHz. To be on the safe side, you should keep a sufficient distance from resonances, e.g. a factor of 10. With this approach, caution is already required at 375 MHz with a 2 cm long piece of floating flooding. Flooding is therefore not recommended for very fast signals [R121]. On the other hand, in an example given in the chapter on power supply design, a device could pass the EMC tests only by flooding. Since much longer unbonded sections may occur, it is essential to control the flooding after it has been performed. It is even better, if possible, to define a design rule for this.

In the past, when ground planes were not available, a "guard trace" was sometimes recommended as a means of preventing crosstalk. A guard trace is a ground line next to the signal track. Guard traces are no longer used today because ground planes are a much better solution due to their much lower impedance.

In addition to keeping the distance to the next potential layer short and using the 3W rule, the following measures can also help prevent crosstalk problems:

- Keep all lines as short as possible.

- Do not parallel signals that do not need to be parallel over long distances.

- Route signals orthogonally in adjacent layers.

- Place components so that the above rules apply.

Mixed-Signal Design Aspects

In circuits that combine digital signals and low-level analog signals—a mixed-signal design—the above rules are not sufficient. The digital and analog circuit components must each be concentrated and placed in separate groups, a process called partitioning. If this grouping and placement is not carefully considered, the result can be a completely non-functional prototype with distorted analog signals.

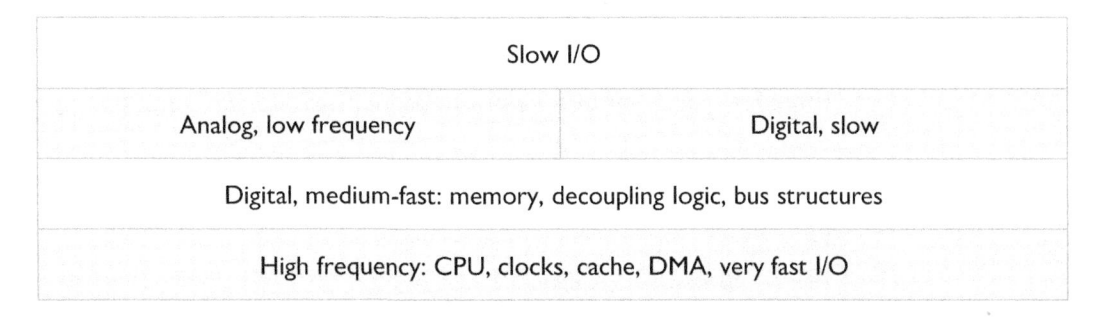

Slow I/O		
Analog, low frequency		Digital, slow
Digital, medium-fast: memory, decoupling logic, bus structures		
High frequency: CPU, clocks, cache, DMA, very fast I/O		

Table 29. Example of a radial migration cascade.

One approach to partitioning is radial migration. We focus on signal speed and create a cascade in terms of bandwidth (see Table 29). It is assumed that a device does not have a wireless interface. Without wireless interfaces, all analog signals are low frequency compared to the fastest digital signals.

The radial migration cascade is mapped onto the organization on the board, e.g., according to Figure 89, by radially following the groups with decreasing bandwidth, starting with the fastest signals.

However, when partitioning according to the radial migration concept, no attention is paid to the following in the meantime:

- How you get signals and power to and from the board.

- Sensitivity to interference. For example, receive circuits are high frequency, but very sensitive to coupled interference, and should not be located next to a clock device.

- How to supply high power areas as directly as possible.

- Distance from DC/DC converter noise.

- Areas with different security requirements.

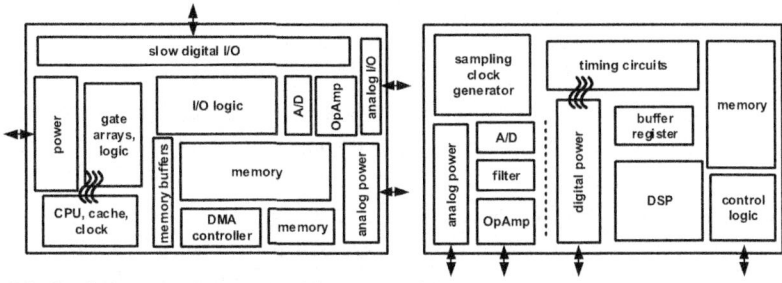

Figure 89. Radial migration cascade examples.

- Distribution of hot components as they heat air and thus indirectly heat other components, e.g. change the resonant frequency, filter frequency, or other parameter of an analog circuit part.

- Grouping elements by power supply (i.e., dividing the board into contiguous, compact areas), each with only one supply voltage.

- Grouping by power loss, where the high-frequency circuits are usually those with the most power loss, at least digital circuits or high-frequency transmitters.

- Place heavy components at the edge of the board where it is attached to the case.

If we look at the partitioning in Figure 89 again, we see the following:

- The power supply is unfavorably solved since it is necessary to supply from two sides. The high-power section has its own power source.

- From a thermal point of view, the placement of the CPU would be incorrect if the card were installed upright with the CPU at the bottom.

- The additional analog power supply allows the generation of a very clean DC voltage with a further focus on low noise, possibly a linear regulator.

The general takeaway from this example is that radial migration is a good first approach. The second step, however, is to check your design against all the other criteria listed above.

No matter how well the partitioning is done, high frequency noise from the digital part will be present in the analog signals. Since the analog bandwidth is usually much smaller, it is possible to use an anti-aliasing filter before the ADC stage. This seems to be a simple matter. Is a simple RC low-pass not enough? Unfortunately, often not. A passive Butterworth filter has an attenuation of 6 dB per octave (per doubled frequency) per pole. To attenuate 60 dB with 1/1000 of the amplitude at the sample frequency, you need at least ten poles, i.e. ten reactive elements such as capacitors or coils. With the help of filter ICs, for example, elliptic filters of high order can be realized with a sharp transition and linear phase response (without distortion in the signal range). This effort can usually be avoided by oversampling (Fig. 93). Single oversampling (i.e., 1-fold oversampling) means sampling at the Nyquist frequency. Double oversampling means sampling at twice the Nyquist frequency, or four times the maximum signal frequency, see Figure 90.

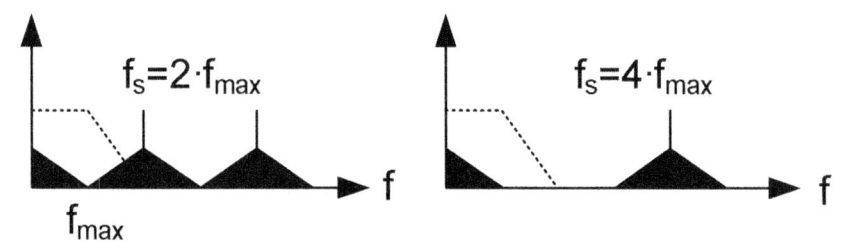

Figure 90. Left: No oversampling. Right: Double oversampling.

Typically, at least 16 times oversampling is used; at 16 times oversampling, the sampling frequency is 32 times the highest signal frequency, with 2 poles we are at 60 dB attenuation, and a second order filter can be easily implemented.

An anti-aliasing filter should also be used with quasi-static signals, such as the voltage from a room temperature sensor. This prevents the influence of noise on the input signal. The cut-off frequency can be set as low as possible. For quasi-static signals, the offset error and the gain error are relevant. However, since the signal is usually post-processed by a microcontroller, these errors can be compensated in software.

Improving the Quality of Digital Signal Connections

Now we move on to the stability of the digital signal connections on the board. A stable digital connection means that it will not spontaneously fail, undisturbed by EMI or crosstalk. Most of the connections on today's PCBs are digital, short, unbranched, direct and finally unproblematic between CMOS ICs: microcontroller to EEPROM, SD card, display driver, touch IC, RAM, recently more and more integrated sensors (IMU, temperature/humidity) and of course communication (USB-UART IC, BLE module/IC). This leads to the temptation to connect CMOS ICs thoughtlessly, "it will work anyway". However, best practices leave nothing to chance; that would be negligent. It follows, first and foremost, that proof of compatibility is required for every connection. We will discuss what that means.

The maximum number of inputs that can be connected to an output of the same logic family to maintain static logic levels is called the (DC) fan-out. For transistor-transistor (TTL) logic, typical fan-out values are in the range of 2 to 10 [R110]. This is a small number, which is why during the heyday of TTL logic (1965–1985) [R111] and long after, this knowledge was highly valued in education. For CMOS, these values are much higher, both for the "old" 4000 family, with e.g. a fan-out of 360 for HEF4000B, and even more for newer families, like e.g. a fan-out of 4800 for the LVC family, see their data sheets. Furthermore, although CMOS is in principle only a technology name, it is linked to certain levels through de facto standardization by the Joint Electron Device

152 Signals on Printed Circuit Boards

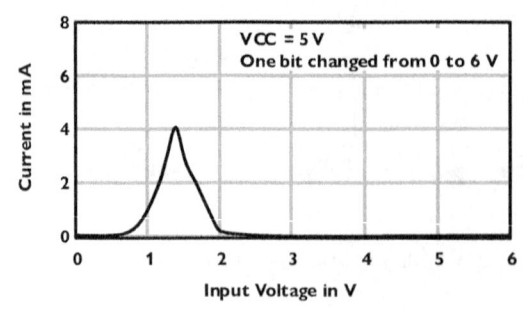

Figure 91. Typical CMOS IC supply current at the specified input voltage [R079].

Engineering Council (JEDEC, also known as the JEDEC Solid State Technology Association). For example, a JEDEC document [R118] lists the well-known 3.3V CMOS levels LOW ≤ 0.8V (24%) and HIGH ≥ 2V (61%).

Thus, for CMOS we can say that the static compatibility of CMOS pins with the same supply voltage can be considered as given, even for many receivers. Since static fan-out is irrelevant for CMOS, "fan-out" for CMOS is sometimes understood as the number of connectable inputs under certain dynamic criteria, such as a maximum edge transition time.

With many receivers on a CMOS output, such as when using a backplane, significant load capacitance can accumulate. Due to the limited output current, the edge transition time can become long. Now, in a suitable situation, one could argue that there is enough time to wait for this transition time. However, this is not a problem.

Figure 91 shows a typical current draw of a CMOS IC when an input without Schmitt trigger function is slowly driven with a voltage ramp. There is a significant current when both transistors are half conducting. This is a cross-current flowing directly from VCC to GND, the so-called "shoot-through current". The resulting power dissipation is the area under the curve ($U \times I$), which is approximately ½ × (2 V–1 V) × 4 mA = 2 mW. This is for one bit; if there are several, the current consumption multiplies, with 8 bits it is already a remarkably high 16 mW. If these bits are switched reasonably often, the average power dissipation is significant. This does not lead to a defect, but to unnecessary

			MIN	MAX	UNIT
		ABT octals		5	
		ABT Widebus™ and Widebus+™		10	
		AHC, AHCT		20	
		FB		10	
$\Delta t/\Delta v$	Input transition rise or fall rate	LVT, LVC, ALVC, ALVT		10	ns/V

Figure 92. Recommended maximum slew rates, to be multiplied by the supply voltage to obtain the maximum slew rate [R079].

heat and wasted energy, especially in portable devices. For logic families, maximum edge transition times are specified in the data sheets (see Figure 92, for example).

In microcontroller datasheets, this information is usually not available, because a microcontroller is by definition not a driver device. You need a driver IC for high capacitive loads.

There is no rule as to whether the inputs of microcontrollers or sensor ICs have Schmitt trigger characteristics. Some microcontrollers have them all, some have them partially, and standard logic ICs have no such inputs at all. Look for "Schmitt trigger" in the data sheet, but also for "hysteresis". In addition, the schematic of the ports must show a Schmitt trigger symbol, if present. Otherwise, assume a normal CMOS input without Schmitt trigger functionality.

The following suggestions can be used to avoid shoot-through current when the capacitive load is high, such as with long cables or many receivers:

- Insert an explicit driver module. The device families for this are 74ABT, 74BCT, 74BTL, and 74GTL. Unidirectional, such as 74ABT244 (octal bus driver). Bidirectional, e.g. 74ABT245, where the direction must be switched (octal bus transceiver).

- For cases with large capacitive loads (e.g. 1000 pF), special high-power drivers are available, e.g. TC4427 with 1.5 A drive current.

- The outputs can also be connected in parallel or a single transistor can be used.

- In such cases, it is better to switch to a more power-efficient signal transfer, such as LVDS, if possible.

Even if the capacity of all connected receivers can be driven fast enough on a backplane, it is worthwhile, especially during development, to add a driver before the signal leaves the board or before it reaches a complex device such as a microcontroller or FPGA from outside. In one project, I had to replace the driver IC several times when a faulty card was plugged into the backplane. This saved me from having to replace the microcontroller.

Most dynamic compatibility problems today arise not because edge transition times are too long, but because they are too short. Colloquially, we talk about short rise times, when what matters is the shortest time, which may well be a fall time. However, I prefer to use the term "rise time" instead of the longer "edge transition time".

Note that we are not discussing clock speed here. Modern CMOS processes result in very fast rise times. These are present even if you clock the system at a comparatively slow clock (e.g. 10 MHz).

In the YouTube video *I only probed the board with a scope—why did my board crash?* Jack Ganssle, a senior electronics engineer, told this story: a customer came to him. They had been making Z80 boards for 30 years, with no clock speed higher than 4 MHz. Once again, the inventory was empty and a new batch was produced. However, none of these new boards worked. The schematic was the same, the layout was the same, and all the component manufacturer names were the same. Eventually, it was discovered that the logic devices that normally switched at 15 ns were now switching at 5 ns. The small, old, slow board suddenly had high-speed signals all over it. Jack had to redesign.

The reason for the faster transition was that the "old" logic device manufacturing had been migrated to a new facility with a smaller feature size (i.e., smaller transistor size), allowing the old facility to be shut down and the new facility to be better utilized. During the migration, the design was ported to the new feature size. However, this resulted in much faster edge rise times.

This is a general problem: for many of today's ICs, a smaller feature size would not be a significant advantage. However, the technical development of fabs is driven by the desire for even faster processors. Eventually, the older, "slower" ICs that still sell well end up there. Thus, over the last few decades, a problem that initially affected only a small circle of high-speed designers, in the case of motherboards, has become a problem that affects all designers.

How to Avoid Overshoot

Since every problematic digital signal has the same basic distortions, the quality of a digital signal can be described by certain characteristics, see also Figure 93:

- Overshoot is the maximum amount of voltage above the setpoint. A negative voltage spike is also an overshoot; to distinguish them, we can speak of "negative overshoot".

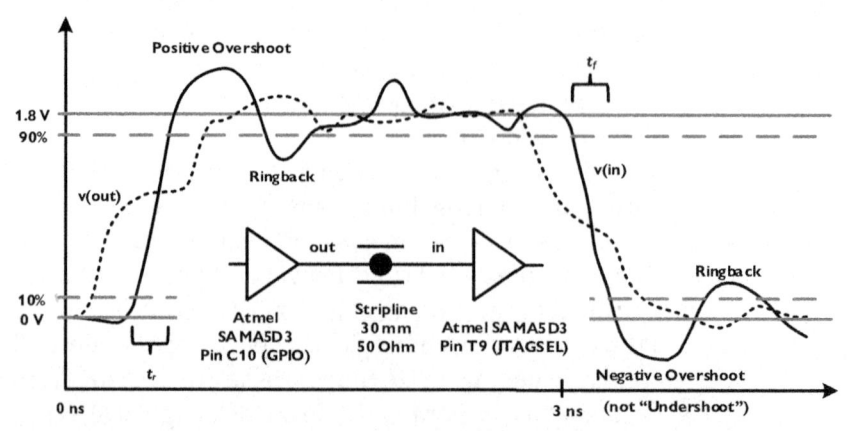

Figure 93. Characteristics of a bad digital signal based on an example.

- The term "undershoot" should not be used. Negative voltage is not undershoot, high-speed guru Lee Ritchey [R126] and Wikipedia agree, but you will find conflicting definitions on the Internet, and even a high-speed course I attended defined negative deflection as undershoot [R125]. Lee Ritchey uses undershoot to refer to ringback and others to refer to the hump in the source voltage (i.e., that the source voltage gradually reaches the target value).

- Ringback is the opposite deflection after the overshoot. A further distinction can be made between "rising edge ringback" and "falling edge ringback."

- Note that "rise time," "fall time," and "(edge) transition time" always mean the shortest time from 10% to 90% or vice versa. This is not consistent in the literature. The transition time for the ADCMP580, a very fast IC, is typically given as 37 ps at "20/80," (i.e., from 20% to 80%). A simple linear conversion is possible with 37 ps × (90–10)/(80–20) = 49.3 ps ≈ 50 ps and shows the still impressive speed of the device.

- The oscillation itself is sometimes called "ringing," but this term originally referred to the oscillation of an LC resonant circuit in response to a pulse.

Regarding the height of the overshoot, it can be stated [R126, R127]: at 3.3 V, the overshoot must not exceed 3.6 V or fall below -0.3 V, otherwise there is a risk of various malfunctions in the CMOS input. At lower supply voltages, the critical voltages are smaller; they correspond to the absolute maximum values. At 5 V they are 5.7 V and -0.7 V (CMOS). Keep a sufficient distance from this limit; it is best not to allow any overshoot at all. Overshoot can occur with any digital connection, not just those with a fast clock. The consequence of the fact that overshoot can occur on any digital link is that each digital signal should be examined at least once during prototyping, unless a special simulation has been run over the layout, more on this below. How to avoid overshoot will be discussed later.

Exceeding the above values is dangerous for several reasons:

- The CMOS structure of N and P field effect transistors (FETs) contains parasitic PN junctions that form a thyristor, also known as a silicon controlled rectifier (SCR). Positive or negative transients can ignite this thyristor. The result is a permanent short-circuit between supply and ground, with the potential to turn off at least the affected input stage [R127].

- Interaction with adjacent inputs through parasitic transistors formed by input protection diodes [R126].

The warning of destruction is not exaggerated. In the same YouTube video *I only Probed the Board with a Scope—Why Did My Board Crash?* mentioned above, Jack Ganssle reports on a similar case he experienced himself. In this case, an ICE with only 6 MHz clock speed and fast edges was tested intensively before the first customer received one. When the customer used the emulator, several ICs on his board "exploded and the IC packages fell apart"!

An overshoot does not necessarily mean destruction. Sometimes it would be better if it did, so that the error would be detected immediately. The following errors are of a "softer" nature and may only occur with certain data, at a certain temperature, etc. The following erratic errors are usually very hard to find:

- An overshoot causes electron injection into the substrate, which causes potential changes in the IC and can cause false triggering of transistors not directly related to the output [R126]. A restart of the system will clear the fault.

- Overshoot is always associated with ringback; if it is too large, false trips may occur as the voltage swings back into the logically undefined range.

- Inter-symbol interference can occur: a new bit change appears at the driver output before the system has settled from the last bit change. The bit changes overlap. This can lead to data loss.

- Timing problems, since a stable digital signal is not present at the next device until after settling.

Therefore, overshoot is undesirable not only for component safety but also for signal integrity reasons. Overshoot is generally avoided whenever possible.

How can you detect potential overload situations as early as possible? In motherboard design, I/O buffer information specification (IBIS) simulations are used as a standard for this purpose. IBIS was introduced by Intel. IBIS files with the extension "ibs" consist of text describing the input and output characteristics of a device. They are purely behavioral and do not reveal the input/output circuitry; intellectual property remains secure. For example, a typical IBIS file for a microcontroller such as TMS320F28069 lists the characteristic for each pin, used as input, used as output, with typical, fastest and slowest rise time, with or without pull-up resistor/pull-down resistor, with or without ESD protection diode to ground or VDD. All of these curves are based on no external load (i.e., no external load capacitance or inductance) and no load resistance, but of course include the internal capacitance of pins [R136, R138]. The curves of the two ICs are used together with their connecting PCB traces to form a signal propagation simulation. IBIS simulations therefore require a drawn layout and are part of a layout program. You can find them in all professional layout programs like Cadence, Mentor Graphics, HPEESof and Altium. There was a freeware implementation called GnuCap for IBIS-3 models around 2013, but no freeware layout program had an interface to it, which essentially killed the project. Neither Eagle nor KiCAD has an IBIS implementation.

If you do not have access to an IBIS simulator, you can still use IBIS data. There are freeware IBIS viewers, or you can use the evaluation version of Micro-Cap, a SPICE simulator that can import IBIS data. When importing, Micro-Cap converts the selected IBIS data into a SPICE model, but it is not disclosed. The component representing such an IBIS source can be found in Version 11 under the menu item "Component, Analog Primitives, IBIS". This allows the most important parameter for potential overshoot—the shortest rising edge time—to be extracted without IBIS simulation. You can also connect a line and a load and simulate the transient. To see the full amplitude of the reflections, only the parasitic input capacitance should be used as a load [R121].

IBIS simulations for motherboards were performed as a check. Signal integrity was already achieved through a sophisticated set of design rules. Without these rules, you would have so many trouble spots that you would have to start the design from scratch.

If you do not have IBIS data, the minimum rise time is often missing as an essential reference for assessing the situation. Rise time data sheets are specified for a certain, usually large, load, such as 15 pF for low-voltage devices or 50 pF for 5V devices. The idea behind this is to guarantee a maximum rise time (see above for the problems of too long rise or fall times). For the actual device, the minimum rise time is lower, since the datasheet specifications always include a safety margin, sometimes up to a factor of 2. For this reason, the edge rise time of the output on a PCB trace cannot usually be read directly from the datasheet. The slower the device, the more difficult it is to find rise time information or IBIS files. For example, if you are using a comparatively small, simple 16-bit, 16MHz MSP430F2012 microcontroller, there is no rise time information in the documents or IBIS models.

$$r = \frac{R_L - R_W}{R_L + R_W} \quad d = 1 + r$$

$$V_r = r \times V_w \quad V_L = d \times V_w$$

Formula 2. Calculation of reflection coefficient r, continuity coefficient d, reflected voltage V_r, voltage at load V_L. Based on characteristic impedance (PCB trace impedance) R_W, load resistance R_L, and incoming signal voltage level V_W.

If information about transition times or simulation models are not available, they can be derived from the clock frequency under certain circumstances. If you do not know the rise time, but know the maximum clock frequency of the device from the data sheet, a good starting point for the rise time is 7% of the period [R089]. This estimate is usually conservative (i.e., the rise time is actually longer). Note, however, that the estimate is valid only if the part has not been ported to a smaller feature size. For devices that are new to the market, this assumption is virtually certain. For those that have been on the market for years ago, the assumption is plausible. For ten-year-old devices, it may not be. For example, according to the datasheet, the MSP430-F2012 was released around 2005. It has a maximum clock speed of 16 MHz and 7% of the period is 4.4 ns. Today, however, it can be manufactured in a process with a smaller feature size that has much steeper edges. For a comparatively slow device like the MSP430F2012, you can just measure its rise time. For fast digital devices, however, you may not have the right, even faster, oscilloscope.

Let us have a little theory to understand the solutions that follow: Any change in AC resistance encountered by a signal causes a portion of the signal to be reflected back in the opposite direction, as described in Formula 2. The reflected part of the signal superimposes itself on the signal and continues to be a combined voltage. With reflections on both sides, settling occurs at the beginning and end of the printed circuit trace.

The transient response is damped by the source and load resistors; the damping from the trace resistance is virtually irrelevant because it is orders of magnitude smaller than the former. If the signal changes slowly compared to the propagation of the signal, any change in the propagating signal is small, and no overshoot will occur. Very fast changes are problematic.

Trace Length Limits

At what line length and signal rise or fall time can overshoot be expected? The acceptable level of overshoot generally depends on the problem (see below). A general rule of thumb is [R121, R139]:

- Transit time < ½ × minimum rise time from 5 V.

- Transit time < ¼ × minimum rise time below 5 V.

$$v = \frac{c_0}{\sqrt{\varepsilon_{r,\text{effective}}}}$$

Formula 3. Speed of propagation of an electromagnetic signal.

The lower value for voltages below 5 V, namely 3.3 V or 1.2 V, is justified by the fact that the interference distances are disproportionately smaller at these voltages. If the fall time is shorter, the fall time value must be used. If the rise time is longer, measures must be taken to reduce the settling. If the rise time is less than 1/6 of the minimum rise time, check the signal with an oscilloscope in the first prototype.

For the above rule of thumb, the propagation time or speed is required. The propagation of an electromagnetic field, and thus of a signal in general, is given by Formula 3. There, c_0 is the speed of light in vacuum, about 3×10^8 m/s and $\varepsilon_{r,\,effective}$ is the effective dielectric constant, i.e., the relative permittivity resulting from the materials around the conductor, as explained immediately below.

Figure 94 shows typical conductor and layer arrangements and typical materials. For the microstrip trace, there are field components in the FR4 material, in the solder mask, and in the air: materials with different dielectric constants and signal velocities.

For a trace width much larger than the height above the next potential layer—the normal case—the field components are concentrated in the material with the highest dielectric constant, so $\varepsilon_{r,\,effective} \approx \varepsilon_r$ of the prepreg material then applies [R105]. If the trace width of a microstrip trace is not much larger than the height above the next potential layer, the dielectric constant is reduced and the signal becomes faster; see R105 for the exact formula.

The same applies to the stripline arrangement. If a potential layer is close, the dielectric constant of the intervening material applies. Since

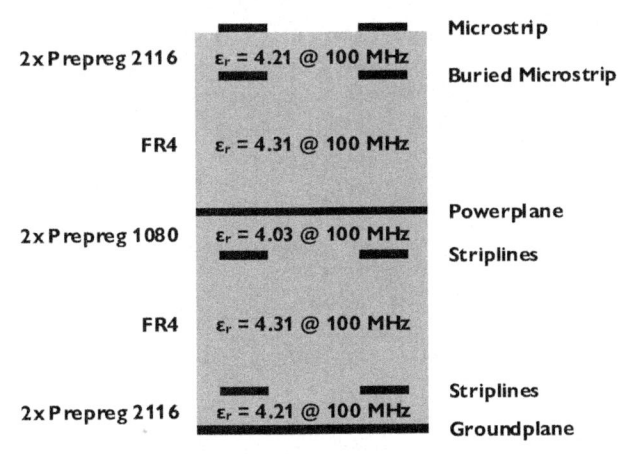

2x Prepreg 2116	$\varepsilon_r = 4.21$ @ 100 MHz	Microstrip / Buried Microstrip
FR4	$\varepsilon_r = 4.31$ @ 100 MHz	
2x Prepreg 1080	$\varepsilon_r = 4.03$ @ 100 MHz	Powerplane / Striplines
FR4	$\varepsilon_r = 4.31$ @ 100 MHz	
2x Prepreg 2116	$\varepsilon_r = 4.21$ @ 100 MHz	Striplines / Groundplane

Figure 94. Microstrip and stripline traces on a 6-layer board. According to R105, traces in the inner layer 2 are also considered (buried) microstrip, but in layer 3 above the ground plane they are considered stripline, since there are no more field components outside the board. In the case of copper areas on both outer layers, all inner conductors are considered striplines, including asymmetrical arrangements and those with adjacent traces.

the power planes are short-circuited to the ground plane via decoupling capacitors, the potential layer does not have to be a ground plane; it can also be a power plane. If the nearest potential layer is far away, as defined above, a mixed formula applies in the same way as for the microstrip trace. However, prepreg and FR4 often have slightly different dielectric constants, so you can just take the average. As a rule of thumb, it is not necessary to know a more precise value to conclude that the following approximation can be used for all structures where the next potential layer is close: the effective dielectric constant is about 4, which means that a signal propagates at about 15 cm/ns on a PCB trace.

The rule of thumb for run-time mentioned above can be rewritten to limit trace length:

- Conductor length < ½ × minimum transition distance from 5 V.

- Conductor length < ¼ × minimum transition distance below 5 V.

The transition distance is the distance traveled in time from a 10% rise to a 90% rise. For the device examples above, and for a generic transition time of 1 ns, these distances are listed in Table 30. Note that we need very short traces to avoid problems with today's fast digital ICs.

Note too that clock speed is not a criterion here at all. For example, R181 reports a situation with a lot of overshoot on a meter of

Part	Rise time t_r (10%-90%) in ps	Transition distance x in cm on PCB with $c_0/2$	Limit length x/4 in cm for VCC < 5 V
AUC Logic	150	2.3	0.6
UltraScale	320	4.8	1.2
SAMA5D3	300	4.5	1.1
ST32L152	500	7.5	1.9
Generic	1000	15.0	3.8
MSP430 F2012	4400	66.0	16.5

Table 30. Rise time and limit trace length. The latter is considered the boundary between low and high speed. However, the signal quality should be checked from a trace length of 1/6 of the transition distance.

CAT-5 twisted-pair cable between a 2.5MHz MSP430F449 and a peripheral card. Also, the trace limit is not the critical length; the later has to do with crosstalk (see earlier in this chapter).

If you did not draw the layout yourself, note on the schematic that the layout designer should give you the length of the longest digital connection.

Signal Trace Impedance and Termination

With today's fast components, even a few centimeters is a long trace. Shortening the trace is not always possible; other solutions are required. In radio circuits, signal reflections on the path from the source to the load are eliminated by keeping the trace impedance constant and selecting source and load impedances equal to the trace impedance. Consequently, as shown in Figure 95, the source must provide twice the nominal voltage as the open circuit voltage. Once connected to a trace, half of the no-load voltage will be on the trace because the internal resistor and the load resistor form a 50% voltage divider. From an overshoot perspective, this is the best solution. However, it is also the solution that wastes 50% of the power in each driver. Obviously, this is not feasible for digital systems with many driver outputs.

What are the other ways to avoid reflections in digital systems besides shortening the trace length?

- Many FPGAs and the GTL/GTLP logic family allow you to select the rise time of the outputs. Do not simply choose the fastest rise time, even if it gives you more timing margin. The correct rise time is a compromise between timing margin and preventing excessive settling.

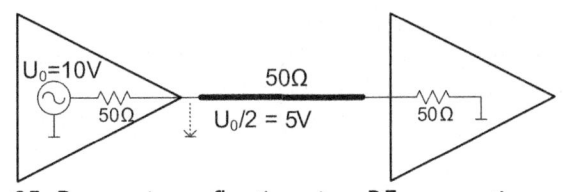

Figure 95. Preventing reflections in a RF connection.

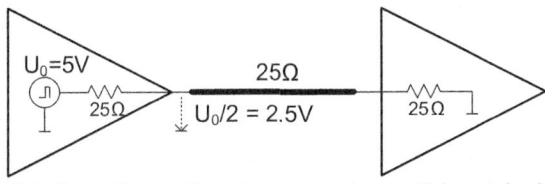

Figure 96. Avoiding reflections is not possible with digital systems as it is with RF connections.

162 Signals on Printed Circuit Boards

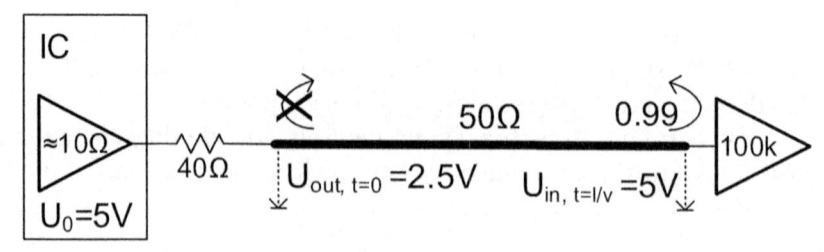

Figure 97. Series Source Termination.

- Replace the IC with one with slower edges.

- In synchronous systems, you may be able to live with reflections if, for example, the clock rate is low and you have time to wait for the signals to stabilize; to be safe, you should be able to wait 5 times the longest run time. Of course, this solution is only possible if the resulting overshoot is tolerable.

If this is not possible, the trace must be terminated. In CMOS systems, series termination, as shown in Figure 97, is clearly the best solution. In this method, a resistor is connected in series with the output so that the sum of the resistor and the source resistance of the driver equals the trace impedance.

Some comments about this solution are as follows:

- The output impedance of the driver must be less than the impedance of the trace for this method to work.

- Series terminators take up space, and matching is only approximate because the driver output impedance is usually not constant.

- First, half of the target voltage is on the trace because of the voltage divider consisting of the trace, the terminating resistor, and the driver's internal resistor. However, the target voltage is still reached at the receiver because of the total reflection at the end.

- Select the series resistor at the lower end of the trace impedance tolerance. This will ensure that the target voltage is reached at the receiver input under all circumstances.

- The maximum allowable distance between the series termination of the resistor and the driver can only be determined by simulation [R121].

- For precise termination, a driver with a defined output impedance, such as the 2245 octal transceiver with 25 Ω source impedance, can be interposed. A terminating resistor is still required to accommodate different trace impedances or deviations from the planned nominal impedance. Logic devices with defined output impedance can be identified by their four-digit numbers.

For the solution shown in Figure 97, the output resistance of the source must be known. In practice, however, this is done by experimenting with the values and choosing one that is the best compromise. Since common ICs do not have a constant source impedance, there is no one correct value because of changes during switching and differences in the LOW/HIGH and HIGH/LOW transitions.

The achievable trace impedance range depends on the geometric conditions. For example, you do not want to realize a very wide trace, and the dielectric thickness cannot be made arbitrarily thin. This means that the lower end of the realizable trace impedance is about 65 Ω for microstrip traces and 40 Ω for stripline traces. On the other hand, a trace cannot be too thin, and the dielectric thickness should not be too wide, otherwise the PCB will simply be too thick. This results in a maximum impedance of about 110 Ω for microstrips and about 80 Ω for striplines.

The higher the trace impedance (i.e., the lower the capacitance to ground), the greater the crosstalk. Therefore, we have typically chosen an impedance at the lower end of the feasible range. For example, 65 Ω has been established for microstrip and 50 Ω for stripline. If you choose 65 Ω for a stripline to allow a reflection-free transition to a microstrip trace, you may be trading yourself for too much crosstalk. Lee Ritchey reports a case where this was done, resulting in a non-functioning backplane. A redesign with a lower stripline impedance was the only solution [R121].

Typical PCBs have a tolerance of ±10% for trace impedance. This corresponds to a range of 45–55 Ω at 50 Ω. Tolerances of ±1.5 Ω may be required, but at a cost. This is because the process remains the same and any PCBs that are out of tolerance are discarded. In general, the cost is too high. Therefore, the circuit is designed from the beginning with 10% tolerance in mind. This is even possible for very high-speed products such as gigabit routers [R121].

However, it is useful to avoid systematic errors. Such errors can occur when switching from one PCB manufacturer to another, or even when your standard PCB manufacturer makes changes, if you do not specify the layer structure very precisely. For example, a 0.1 mm thick prepreg can be made with laminate 2113 and 54.4% resin, resulting in a dielectric constant of 4.0 at 1 GHz. A prepreg of the same thickness made with laminate 2116 and 43% resin will have a dielectric constant of 4.37 at 1 GHz [R121].

There are three ways to design an impedance controlled trace:

- The PCB manufacturer provides you with a program that you can use to simulate the routing. The program contains manufacturer-specific data, such as layer and copper thicknesses and their tolerances. At best, this is a 2-D field simulator; at worst, the results are approximations.

- You get a collection of approximation formulas.

- You get tables of typical line types and their impedances.

There are different opinions about these methods:

- Manufacturers supplying the calculation programs point out the coverage of all possible conductor arrangements.

- Manufacturers promoting impedance tables point to ease of use with guaranteed accuracy.

Calculations using approximation formulas give less accurate results than field simulations because the formulas are fitted to the results of the simulations. When using approximation formulas, it is important to know their range of validity. Programs that use approximate formulas ensure this by allowing only valid dimension choices. If you use a formula by hand, you are responsible for it. Therefore, this is the least safe form, followed by a program that has only approximate formulas in the background. The best method is still a field simulation using a tool recommended by the manufacturer and the same material specifications. It is important to have accurate values for the dielectric constant of the prepreg layers in the frequency range of the application. Values at 1 MHz, as usually found in data sheets, are not sufficient. Most of the energy of a signal is in the first harmonic of the switching edge; this is calculated as $f = 0.35/t_r$ with a 10% to 90% rise or fall time t_r, whichever is shorter. The dielectric constant at this frequency must be known. However, the actual value may be different if the PCB manufacturer uses a different glass/resin ratio. In any case, you must identify a manufacturer early and discuss the design of impedance controlled traces.

You should place a test coupon on the board or in the center of the panel of panelized boards to verify successful impedance realization. This is usually suggested by the manufacturers themselves to verify successful fabrication. The coupon must be at least six inches (15 cm) long for standard test equipment and have a connector pattern at its ends that matches the test equipment. This pattern may be several centimeters wide if you need test coupons on all layers, because the coupons must be spaced far enough apart laterally. There is usually not enough space on the board itself for all of them, and as mentioned above, the test coupons are placed between the panelized boards.

It is important to agree on an edge slope as an integral part of the coupon impedance test. The measured impedance can differ by several

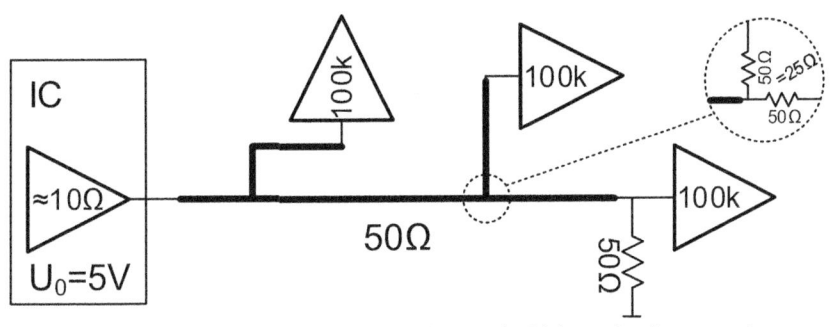

Figure 98. Failed distribution of a fast digital signal: Although all traces have an impedance of 50 Ω and there is a suitable termination at the end of the trace, there are intermediate reflections at the branching points.

ohms depending on whether a 40 ps slope or the more common 175 ps slope is used, since the dielectric constant changes with frequency.

The manufacturer may retreat to guaranteeing only mechanical correctness (e.g., layer spacing, copper thickness, and trace width within tolerance), but you should not accept this.

The outer layers allow only microstrip traces. Their impedance is subject to greater manufacturing tolerances than traces in the inner layers because the copper plating on the via wall also includes the copper plating on the outer layers. The risk of excessive emission and/or crosstalk between the two potential planes is greater than for inner layer traces. It is recommended that outer layers are only used for non-critical signals [R121].

Microstrips in the second or deeper layer with a single potential layer nearby, called buried microstrips, are equivalent to striplines. By keeping the height above the potential layer short (e.g., 0.2 mm), there is no increased emission or greater crosstalk with microstrips than with striplines.

Striplines do not need to be limited by the potential layers in both adjacent layers. It is more important that one of the potential layers be as close as possible to the stripline (e.g., 0.2 mm). If there are two signal layers between the two potential layers, the traces should be placed orthogonally.

Branching at Fast Digital Signals, Parallel Termination

Ill-conceived long branches, even with an impedance-controlled trace and proper termination, will result in a mixture of signals and their reflections. Each branch has a different impedance to the incoming signal than the trace itself (see Figure 98).

However, parallel termination works quite well with sequential taps, as shown in Figure 99, under the following conditions:

- The branches or stubs must be electrically short. R059 specifies a maximum length ≤ 1/10 of the edge transition length, i.e. the length covered during the fastest edge transition. R121, on the other hand, allows a maximum length of ⅓ of the edge transition length. If possible, the stubs should have a lower capacitance per length than the main trace.

- If the stubs are too long, this will cause stub ringing. One way to counteract this is to insert a resistor at the tap point. This is done with SIMM sockets. The resistor is followed by a 50 Ω impedance-controlled trace to the input. This solution works up to about 200 MB/s per trace [R121].

The disadvantages of this solution are as follows:

- The signal does not reach the components at exactly the same time; this may not be tolerable for the clock.

- The input capacitance changes the impedance of the track.

- Since parallel termination converts a lot of power, it is typically operated at the lowest possible high-level voltage. However, this results in reduced noise immunity. In general, therefore, parallel-terminated connections are more susceptible to problems with power supply noise.

- As soon as the length from the connection pin in the IC to the effective gate of the receiver transistor becomes too long, stubs can no longer be used. This is the case below 300 ps edge rise time. Multidrop traces cannot be implemented for gigabit links.

For parallel termination, the output impedance of the driver must be sufficiently low, preferably a factor of 10 lower than the trace impedance [R121]. This is because the trace forms a voltage divider with the internal resistance of the driver. If the internal resistance of the driver is too high, a valid high level will not be reached, even if the termination is correct. Select the termination resistor at the upper limit of the

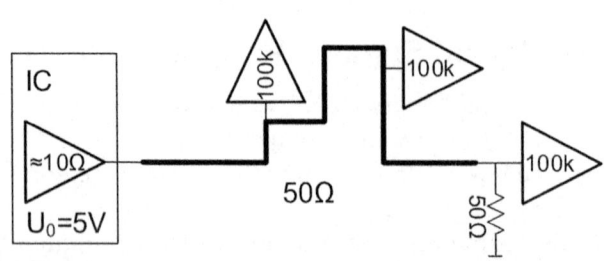

Figure 99. Stubs are possible on a parallel termination as long as they are short. Even the last piece can be a stub if you have no space directly at the pin (e.g. on a BGA IC). This is called "terminator end swapping".

trace impedance tolerance so that reflections caused by a deviating trace impedance always remain positive, increasing rather than decreasing the observed high-level voltage.

Sometimes a mismatched termination is made. This results in a certain reflection coefficient for a higher load impedance.

There are alternatives to terminating the resistor to ground, but they all have drawbacks. AC termination can be considered (i.e., a capacitor in series with the terminating resistor). The capacitor avoids a high continuous current at a constant high level. In practice, this termination proves to be problematic. The selected capacitor must not be too small, otherwise the overshoot will be too high. If a large enough capacitor is chosen, the digital signal becomes sinusoidal and the termination acts as a low-pass filter, although the signal is not tapped between the resistor and the capacitor. The additional capacitor takes up space and costs money. The capacitor is not ideal and does not work in the intended sense in the filter up to arbitrarily high frequencies. Therefore, AC termination is not recommended [R121].

Terminating with diodes to ground and supply introduces current pulses to the supply and ground (i.e., it increases supply and ground noise). The supply noise can be reduced with decoupling capacitors, but you cannot do anything about the ground noise. Therefore, it is not a suitable termination method. It is also the most expensive type of termination. If you decide to use it, you must use very fast Schottky diodes. The internal ESD protection diodes are too slow, otherwise we would never need to terminate.

A Thevenin termination (i.e., an additional resistor to the supply voltage) can improve the situation if one edge of the driver is related to a lower source impedance. This is the case with TTL, where the falling edge has a faster transition time than the rising edge. Using a 220 Ω to 330 Ω Thevenin voltage divider to raise the LOW potential from 0 V to 3 V helps the driver on the rising edge. However, this worsens the transition time on the falling edge. Overall, both edges become comparably fast. It should be noted that such traces must always be driven, otherwise a voltage in a logically undefined range will be present at the receiver, which can be associated with a constantly high shoot-through current. Thevenin termination is required when the LOW potential does not correspond to ground. An example is LOW-voltage positive emitter-coupled logic with 1.6 V for LOW. The voltage divider of the Thevenin termination is used to set the LOW level. The impedance of the termination is given by the parallel connection of the two resistors.

Taps and stubs change the impedance of the main trace. As shown in the example in Figure 100, the change can be significant. However, you can easily adjust the terminating resistor.

168 Signals on Printed Circuit Boards

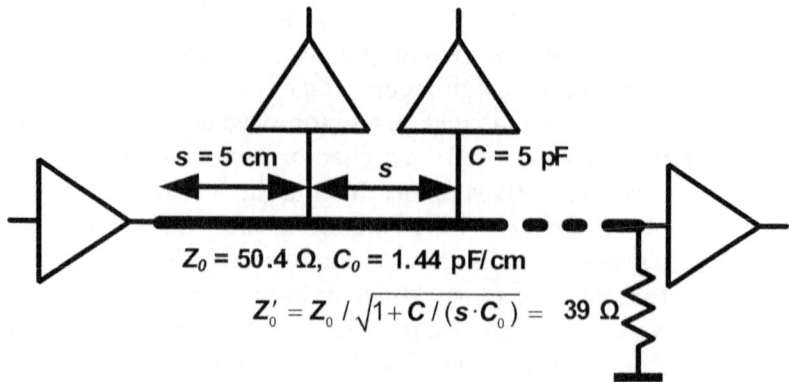

$$Z_0 = 50.4\ \Omega,\ C_0 = 1.44\ \text{pF/cm}$$
$$Z_0' = Z_0 / \sqrt{1 + C/(s \cdot C_0)} = 39\ \Omega$$

Figure 100. Taps and stubs change the impedance of the transmission trace. In the example shown, a 5 pF stub is added every 5 cm, which is short enough according to the text. This changes a 50.4 Ω transmission trace to one with 39 Ω impedance, so the terminating resistor must be adjusted.

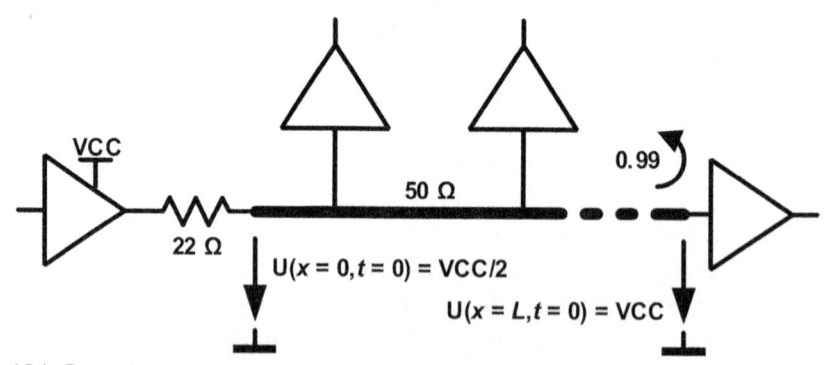

$$U(x = 0, t = 0) = VCC/2$$
$$U(x = L, t = 0) = VCC$$

Figure 101. Branching and series termination are problematic.

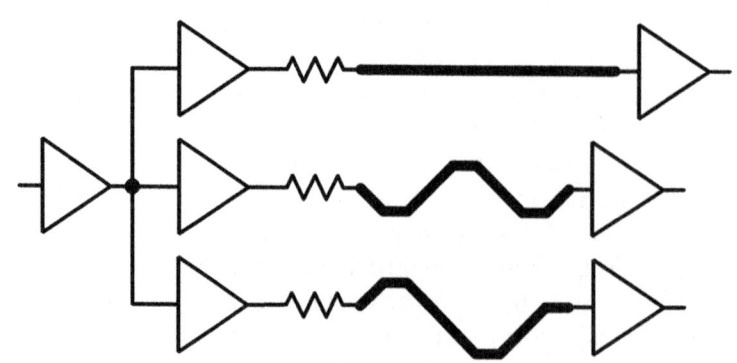

Figure 102. Clock distribution with series termination, star connection, and meanders.

Unidirectional Signal Series Termination

The following problems occur when attempting a series termination of a multidrop trace (Figure 101):

- The first element may have to wait a long time for the input level to become valid, since it is created after the total reflection at the end of the trace.

- As long as the signal travels forward on the main trace, its level is in the logical intermediate range, with the risk of increased "shoot-through" current.

However, this type of termination is used in practice with the PCI bus. A PCI bus always has a limited bandwidth. The 100 MHz PCI bus has a clock period of only 10 ns. This results in a maximum bus length of 12.7 cm. Within this time, the clock edge has oscillated back and forth to the driver before the new clock edge arrives.

A clock signal usually requires simultaneous arrival at the destination. This is only possible with a star-shaped distribution. This allows a low-loss series termination. Of course, all trace sections must be of equal length. This usually requires meanders, as shown in Figure 102. These structures seem to resemble a coil, so you may wonder if it will work at all. However, experience shows [R105] that meandering is trouble-free when using a stripline. The meander is placed in a dedicated signal layer and is not embedded in a potential plane.

There is disagreement about the importance of a constant cross-section (i.e., rounded or at least 45° cut edges); see below for the 90° angle issue. The crosstalk between the individual meander turns is only relevant at GHz frequencies. There is no consensus on whether a via length must be added to the total length when the trace goes over a via (R0105).

Termination Bidirectional Connections

A bidirectional connection between two ICs can be easily terminated in series (see Figure 103). A bidirectional bus with more than two nodes can be terminated in series like the PCI bus, for example, but then the bandwidth limitation mentioned above applies.

Figure 103. Bidirectional connection termination.

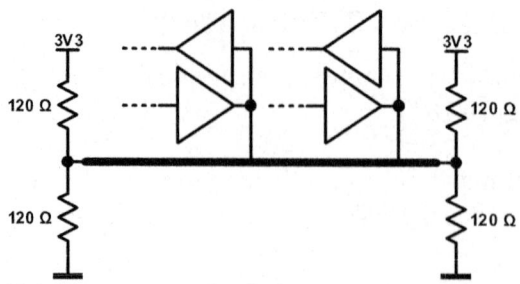

Figure 104. Termination for bidirectional connection with multiple nodes.

A bidirectional connection with multiple nodes without bandwidth limitation is only possible by using a parallel termination (see Figure 104) with a Thevenin termination. Such a termination on a backplane is necessary, for example, if you do not know in advance which slot cards will be inserted. The disadvantage of backplane termination is that it must be populated with SMD components. If you knew for sure which cards would be in the first and last slots, you could terminate on the cards.

The Stratix III and IV FPGAs offer a mixed variant for a bidirectional connection with multiple devices: You place them at one end of the trace and can then use dynamic on-chip termination (dynamic OCT). This automatically performs series termination on transmit and parallel termination on receive without the need for additional termination resistors. This is similar to the dynamic on-die termination (ODT) feature in DDR3 RAM devices, where termination can also be switched based on activity.

Terminating a Bidirectional Bus

On a bus with parallel tracks, the impedance of each track also depends on the potential of the adjacent tracks when the bus is on an outside layer. The magnitude of this change is called the impedance swing. A large impedance swing makes it difficult to terminate the traces well.

One solution is to embed the bus in an internal layer. Use the 2H rule and place the bus between a ground plane and a power plane. This is typically done with backplanes, creating a stripline configuration that virtually eliminates the impedance swing.

Alternatively, use the 3W rule. A large impedance swing always means high crosstalk, and vice versa. If the crosstalk is eliminated by applying the 3W rule, the impedance swing also disappears. However, the space required is considerable.

Differential Signal Routing Details

All of the above considerations have been limited to unbalanced ("single-ended") connections. These connections produce noise on the supply and on the ground. The noise on the supply can be minimized to some extent by using decoupling capacitors and a power-ground plane capacitor. There is no such antidote for ground noise. Since the HIGH level voltage has to be reduced with increasing switching frequency to avoid too much power dissipation, a limit is reached at about 200 Mbps, beyond which one no longer gets monotonous rising and falling edges with an asymmetrical connection. The noise on the ground is transferred 1:1 to the digital signal, and at this limit it has reached such a strength that it alters the edges beyond permissible limits. Therefore, a differential connection must be used for higher speeds. Since both traces of a differential pair carry the ground noise as common mode noise, it is eliminated at the differential receiver and we get a clean edge again.

Figure 105 shows an example of a data bus between a processor and double data rate random access memory (DDR RAM) operating at 1.35 V. At this voltage and a rise time of 300 ps, the trace length limit is 1.1 cm, as shown in Table 34 above. Since the data bus is longer, it must be terminated. In principle, this could be done with series resistors on both sides, but there is not enough space. Instead, it is terminated internally in the ICs based on the ohmic channel resistance of

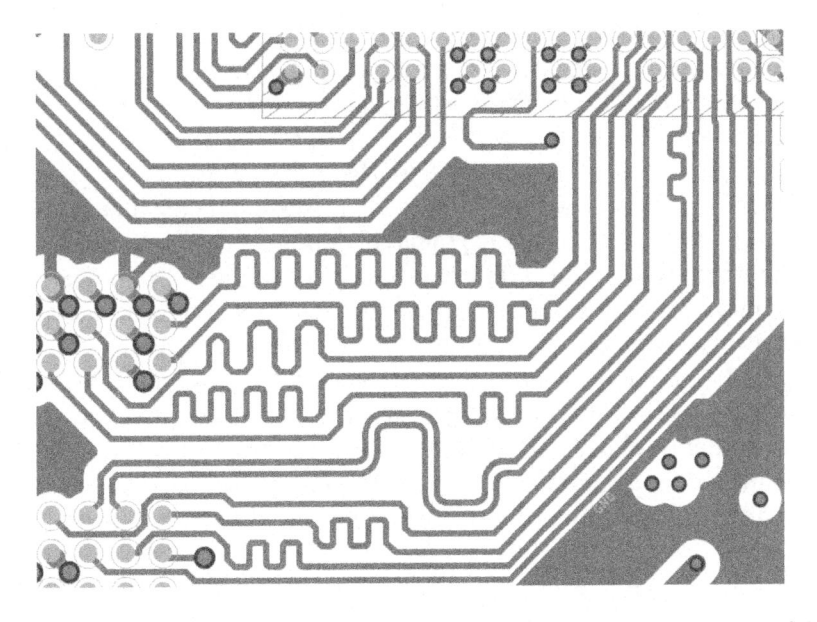

Figure 105. Section of the connections on the top layer between the ARM9 microprocessor (top, right) and a DDR3 SDRAM (bottom, left) of the i.MX23-OLinuXino board. The widely spaced traces are data traces, and the two closely spaced traces are data strobes with a length of 20 mm. The rise time of the 454 MHz processor and the 130 MHz memory is in the range of 300 ps.

a MOSFET, a so-called on-die termination. The available termination resistors for DDR3 RAMs are 120 Ω, 60 Ω and 40 Ω. The trace impedance must be designed accordingly.

The data rate in the example shown in Figure 105 is 260 Mbps because the DDR memory uses both the rising and falling edges for data transfer. However, notice that only two traces are symmetrically routed. These traces carry the data strobe signal. Its edge determines the valid time of the data. Therefore, the data traces themselves do not need to be symmetrically routed because they are at a safe level when the strobe signal arrives.

The data traces are therefore still asymmetrical. A 3W distance is maintained to prevent significant crosstalk. Meanders are used to equalize the propagation time of all bits on the bus. The resulting nearly right angles are discussed in the next section.

The data strobe pair is conspicuously close together. This is not necessary to eliminate ground noise. The reason for this could be the approach to make externally coupled or radiated noise appear as common mode if possible. The extent to which this is useful depends entirely on the sources of interference on the board and in the environment. This approach also has a drawback. The result of tightly routed differential traces is destructive interference with flattened signal edges and reduced impedance of the individual traces, so they must be made narrower to get back to the originally planned impedance. For this reason, some people believe that differential traces should have a minimum distance between them. If you follow the 3W rule, the coupling between the traces of the differential pair is so small that we can consider them as separate traces. The two traces in the pair can then be kept completely separate, which is of course a huge advantage in unbundling. Moreover, their lengths do not need to match exactly. According to the calculations and experience of a designer of Gigahertz routers, length differences of up to 7.6 mm are permissible at 2.4 Gbps [R141]. This allows one trace to have more vias than the other, giving the autorouter the necessary freedom in a complex design. This result also means that it is pointless to debate whether differential traces are better placed coplanar or on top of each other ("broadside").

Differential unbranched connections can be terminated separately using one of the available single-ended termination schemes. This means that both traces do not necessarily have to have the same impedance. Using a parallel termination scheme, such as on a backplane, the differential termination shown in Figure 106 can save one resistor.

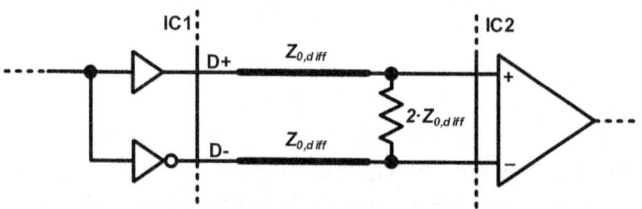

Figure 106. Terminate a differential trace pair.

Tracks at Right Angles

One of the most persistent rules of thumb for routing fast connections is that you should not draw any right angles, but that each arc should have at least 45° angles. It was shown in 1998 that this does not matter at all [R190]. Even with an edge transition of only 125 ps, no influence of the implemented right angle was detected in the experiment. A possible source for the idea to avoid right angles in tracks was a diagram in the *Motorola ECL Systems Handbook* from 1973, which showed the effect of a 90° angle on the signal. It turned out, however, that the dips shown there were not caused by the 90° angle at all, but by something else; the measurement was simply wrong, as the author later admitted. A right angle is not a cause of increased radiation. Acute angles have long been a problem in PCB manufacturing. When PCBs were screen-printed, ink sometimes flowed into an acute angle; this had no electrical consequence at the time, but it gave the impression of an error and the layout had to be revised. Finally, PCBs have long since been washed in such a way that acids cannot remain at acute angles.

Vias and Fast Signals

A similar mantra regarding right angles and traces holds true for signal path vias: it is impermissible to route fast signals on impedance-controlled traces through vias. On the other hand, it has already been shown in the chapter on power supply design that the inductance and resistance of a single via are negligible, even for gigabit/s signals. In addition, the via capacitance must be taken into account. This was not the case with the supply and ground traces. The via capacitance results from the via breaking through the power and ground planes (i.e., the via ring represents the dielectric). This results in capacitance, but it is small. R121 has an average capacitance of 0.25 pF for a 0.3 mm diameter via on a 4-layer board. Together with the typical channel resistance of an NMOS transistor of 10 Ω, I get a time constant of 2.5 ps. It is safe to say that the influence of the via on the signal is negligible.

The view of impermissibility may stem from the fact that the use of a via involves a principal hazard. With the layer change, there is also a change to a different track. For an ordinary PCB with a 10% manufacturing tolerance to the target impedance, this means a maximum reflection factor of -4.8% or +5.3% at 50 Ω. Basically, this is not ideal, but since you have to design the connection for the tolerances of the trace impedance, the terminating resistor, and the driver, this additional tolerance does not matter. It follows that you can safely place at least one via in a high-speed trace. If there are several larger ones, such as two press-fit connectors on a backplane, the situation must be examined from a measurement point of view.

Tri-State Considerations

We now turn to a special case that also belongs in this chapter. This example is shown in Figure 107: two drivers with tri-state outputs can apparently be connected if one driver's output is switched to high impedance. However, the following should be noted:

- A "floating" ("tristate") switched output will not automatically tolerate a voltage higher than the operating voltage of the device. The example in Figure 107 may not work. In most cases, you will not find this information in the datasheets. The best practice is to connect only ICs with the same supply voltage level.

- If all controlling components are set to "tristate," there may be no logically defined level at the receiver module. The leakage currents will result in a certain stable voltage ("floating state") that may be in the logically undefined range.

One solution to the last problem is to use a pull-up or pull-down resistor. However, the resistor forms a low-pass filter with the input capacitance and the trace capacitance, i.e. it can slow down the rise time. The appropriate resistor value is calculated as R = (maximum allowable rise time at the input) / $(0.4 \times C)$ where C = the sum of the input and trace capacitances. The input capacitances can be found in the data sheet; a value of 0.5 pF/cm can be assumed for the trace capacitance.

Another solution to the "floating state" problem is to use components with "bus hold" property, see Figure 108 for an explanation of this concept. Multiple bus-hold ICs can be placed on the same bus. A bus-hold device can be thought of as a dynamic terminating resistor and has similar results to a terminating resistor in terms of dynamic losses and reflections.

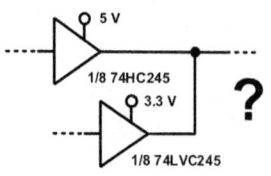

Figure 107. Connecting drivers with tri-state outputs.

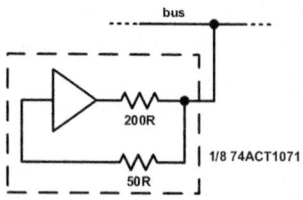

Figure 108. Schematic of a bus hold circuit inside the 74ACT1071. It is a driver with a relatively high source impedance to be easily overridden, and with feedback from the output to the input that maintains the last externally driven state.

5 Thermal Management

A study by the U.S. Air Force Avionic Integrity Program found that 54% of premature electronic equipment failures were due to overheating [R029]. Thermal management can be challenging. Time and again, I have seen cases where a major redesign of the board and package late in the design process became unavoidable; otherwise, the device would simply overheat. In many of these cases, however, the engineers admittedly did not pay enough attention to the thermal aspects at the beginning of the new design. The lesson to be learned is that the thermal management of a new design should not be postponed, underestimated, or simply brushed aside. This chapter is dedicated to raising awareness of this issue and discussing measures to prevent overheated components. We will also briefly discuss the challenges at the other end of the temperature spectrum, at and below 0°C.

Research on semiconductor failure rates [R002, R029] has yielded an impressive result: the lifetime of a semiconductor device is halved for every 10°C increase in temperature. This finding is consistent with the Arrhenius equation, often cited in this context, which states that chemical reactions are accelerated by a factor of two to four when the temperature is increased by 10 K [R166]. The operating die temperature of a semiconductor is therefore the most influential factor in its lifetime. Therefore, it is not a matter of landing precisely a few degrees below the usual maximum semiconductor temperature of 150°C specified in the data sheet. For example, the U.S. military requires a junction temperature limit of 110°C for operating semiconductors [R161]. Texas Instruments defines a 10-year lifetime for its SM320F28335 high-temperature microcontroller in a plastic package at a die temperature of 105°C. In contrast, semiconductors can operate well beyond the 150°C junction temperature. In a ceramic package, the same microcontroller is specified up to a 175°C junction temperature. However, at this high temperature, the guaranteed lifetime is only 500 hours. The common upper limit of 150°C for the junction temperature is not due to the semiconductor material itself, but to the plastic packaging. At a continuous package temperature of about 165°C, outgassing causes polymer additives, such as flame retardants, to migrate from the package to the semiconductor and degrade performance. In addition, loss of plasticizer causes the package to become brittle and fragile.

Caution Defining the Ambient Temperature Range

Avoiding thermal problems starts at the specification stage. Semiconductors that do not dissipate significant power, such as EEPROMs, exist in specific ambient temperature ranges. Specified temperatures for industrial-grade components typically range from -40°C to +85°C, and sometimes as low as -25°C. Commercial components are typically specified from 0°C to 70°C, and military components are typically specified from -55°C to 125°C. Automotive components often range from -40°C to 125°C. As noted, these are ambient temperatures. A powerful microcontroller running at full speed in a sealed enclosure can easily result in an internal temperature of 20°C above ambient. Therefore, you should try to set the defined upper limit for the ambient temperature of a device as low as possible.

The lower limit is also problematic. Thermal shock (i.e., bringing the device from the cold outside into a heated room) will cause condensation on the PCB unless it is in a completely sealed enclosure. Water on the PCB can cause galvanic corrosion, with pin-to-pin resistances dropping below 1 MΩ. In high-voltage circuits, this results in inadequate isolation. The next problem occurs when the device is returned to an outside temperature below 0°C. The dew freezes, causing mechanical stress between the pins of the IC and the PCB. Therefore, if possible, the ambient temperature you specify should not go below, e.g., 4°C.

If the ambient temperature is very low, new problems arise. Below -20°C, an LCD may be slow to respond or even freeze [R167]. An OLED display is a solution here; it works down to -40°C. At -30°C, most aluminum electrolytic capacitors freeze. Tantalum solid electrolytic capacitors are recommended as an alternative for operation below -25°C. Experience shows that the mechanical properties of connectors and switches are also different at low temperatures and must be tested.

It should also be noted that a wide allowable ambient temperature range includes a corresponding allowable temperature shock. If possible, a maximum temperature gradient should be defined that matches the heating or cooling rate of the oven and climate chamber used in the temperature tests. If this is not possible, thermal shock testing must be performed. Extreme thermal shock can tear PCBs apart due to different mechanical expansion coefficients.

Outside the ambient temperature range for the specified component, re-specification by the manufacturer is required, if the manufacturer addresses it at all. You cannot rely on a batch you have tested. The manufacturer can always make changes that will not affect the performance of the device in the specified temperature range but may affect it outside. There is no rule about what happens to the function of the component at temperatures outside the specified range. At extremely low temperatures, a transistor may perform worse or even better, it may suddenly stop working or gradually deteriorate, or it may become unstable and start oscillating [R166, R167].

Common and Unusual Suspects That Get Hot

In power electronics, the need for thermal management is obvious. This does not seem to be the case in signal processing. However, today's microcontrollers and FPGAs can easily dissipate a few watts at full load. A powerful microcontroller can be the hottest component on a signal processing board, not the DC/DC converter. If you are aware of this issue, you will want to study the thermal data in the microcontroller or FPGA datasheet. It should be noted that the usual R_{thjc} value (i.e., the value of the thermal resistance from the junction to the package) is useless if the IC has a thermal pad on the bottom side. After soldering to the board, this pad will be inaccessible, but it will be much hotter than the top of the package. This is why you will find the value of "junction-to-package top" in any serious datasheet. This quantity relates the junction temperature to the temperature at the center of the top of the IC [R165]. This is not the thermal resistance of the entire top surface because it is heated unevenly. For this reason, the value is denoted not by R but by the Greek symbol Ψ (Psi). Multiplying this value by the dissipated power and adding it to the measured temperature at the center of the top of the IC gives a good estimate of the junction temperature. This is still an estimate because the actual value depends on how well the heat is dissipated at the bottom of the device. On a four-layer board with two signal and two potential layers (2s2p), if you connect the heat pad to a ground plane through thermal vias, the value of Ψ is about half that of a single-sided (1s0p) board with no ground or supply plane [R160]. Airflow increases Ψ because more heat is removed from the top surface. Check the conditions for the Ψ values given in the datasheet.

Estimate of the Internal Case Temperature

Signal processing electronics are often housed in sealed enclosures. Figure 109 shows an approximation of the temperature rise inside such a housing. The curves in the diagram are based on evenly distributed heat generation inside the enclosure. How this can be achieved is discussed below. If this assumption is correct, the so-called free surface of the enclosure has the greatest influence on the temperature rise inside [R003]. A free surface is a side of the housing facing the ambient air. Mounted sides are not free, meaning the temperature to which they are exposed is unknown. Consider a device mounted on the outside of a metal cabinet. Initially, the mounted side sees the ambient temperature, assuming there are no heat sources in the cabinet. However, the part of the cabinet to which the device is attached to gradually becomes hotter and is heated by the device—the cabinet has good but limited thermal conductivity. The result is a temperature that is higher than the ambient temperature, but the increase is unknown. It makes sense to exclude this side from the calculations altogether.

178 Thermal Management

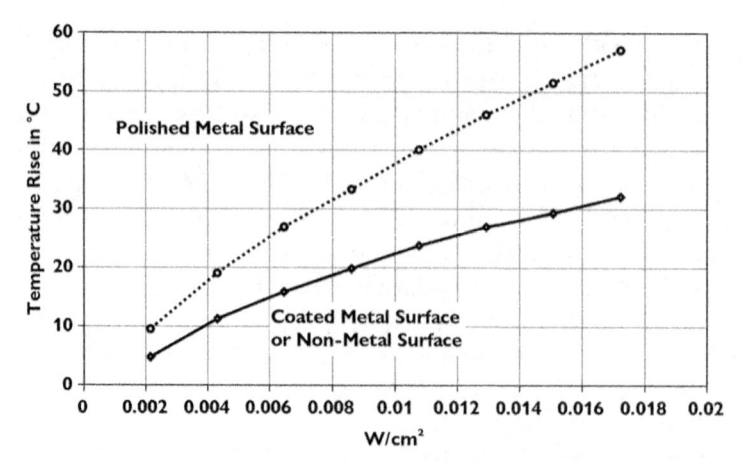

Figure 109. Free surface and internal heat rise above ambient, depending on the heat generated in the device and the surface treatment [R003]. The two measuring points on the far left correspond to a cube with an edge length of 10 cm, five free surfaces and a heat dissipation of 1 W inside.

In addition, the surface finish of the enclosure has an important influence. If the housing is painted, the visible color does not matter because all non-metallic coatings have an emissivity within 0.85–0.95 [R145, R146]. The thickness of the coating is also irrelevant if it is thicker than 40 µm [R147]. A standard triple spray coating has a thickness of 25 µm to 127 µm, and a less common single spray coating has a thickness of about 12 µm [R148]. Anodizing is similar to painting, and the layer can be so thin that it appears transparent.

Polished metal surfaces tarnish or oxidize on their own, both of which improve emissions and reduce the temperature rise to the midpoint between the two curves in Figure 109 [R146].

For example, consider a PCB with an area of 100 x 80 mm² housed in a plastic case of approximately the same length and width. Let the height of the case be 3 cm. Then, for a dissipated power of only 1 W, we get a temperature rise of over 10°C compared to the ambient temperature. A margin of +25% should be used for such a value obtained from Figure 109, since the heat inside is most likely not evenly distributed [R003]. The use of circulating fans in the cabinet reduces this margin to about +10%. Distributed placement of multiple hot devices also reduces the required margin.

The curves in Figure 109 are not applicable if the unit is also heated from the outside. This is especially the case if it is exposed to full sunlight. It is known from the literature that the internal temperature of an enclosure in Central Europe can reach 70°C without internal heat dissipation. In deserts, temperatures above 90°C have been reported [R019, R026]. Nissan specifies 81°C for electronic components in a car and a maximum surface temperature of 111°C for the instrument panel [R166]. Tests are mandatory and must be conducted in the absence of wind, as this has a significant effect on the results.

Since most of the sun's radiation is in the visible wavelength range, light-colored or reflective surfaces are best for reducing solar heating. A white or metallic enclosure color can reduce the temperature rise to one-fifth that of an optically black enclosure [R003]. Using a top sun-shield just a few inches above the unit can reduce the internal temperature by 25%, and top and side shielding can reduce it by 45%.

As mentioned above, the visible color of the device is irrelevant for heat dissipation. For satellites, for example, Teflon™-coated silver foil is used. The thin Teflon layer is transparent, and the underlying silver foil completely reflects sunlight. In contrast, the Teflon layer efficiently radiates the heat generated in the satellite because it has a high thermal emissivity [R149].

Placing Hot Components in a Closed Case

In a sealed plastic enclosure, the hot components should be in as direct contact as possible with the enclosure [R016]. Even a small air gap will significantly increase the temperature of the hot components. Distribute hot components (e.g., transistors in TO-220 packages) along the sides and microcontrollers toward the lid. Fill the remaining space with soft, thermally conductive materials such as thermal rubber. Consider using thermal foil on the inside of the enclosure wall with a thermal conductivity greater than 50 W/(m×°C) [R029]. Up to 30 W, a plastic enclosure can perform as well as a metal enclosure in terms of heat transfer [R016]. A large PCB heat sink and its surrounding space will generally be worse.

If the case is too large to be filled and the hot components cannot be placed on one side or on top, use a fan to distribute the heat.

Unlike metal cases, it is not important to choose extra thick walls or to add a plastic heat sink to the outside of the case [R016]. Note that the maximum contact temperature is not exceeded for all materials, as mentioned below.

Placing a Heat Sink Outside the Case

An obvious variant that has not yet been discussed is the possibility to mount a heatsink outside the case. A hole in the case allows you to mount a transistor directly on it, without a case wall in between. As elegant as this may seem at first glance, its implementation becomes complicated and problematic. The solution creates a connection between a component soldered to the PCB and a component mounted on the outside of the package. This results in a complex assembly process that must be discussed with the manufacturer in advance, as it may end up being too expensive to implement. If you choose this option, be aware of the maximum contact temperature as shown below. Also note that the heat sink must be electrically connected to the case if it is made of metal. This is not only for safety reasons if the device is powered from the mains, but also to prevent ESD from jumping from a

180 Thermal Management

floating heatsink to the PCB. This requires the controlled use of three metal materials in contact—the heat sink, the screws, and the metal enclosure—so that the electrical connection is not disabled by corrosion. Using the metal case itself as a heat sink will most likely not work because it would create hot spots with excessive contact temperatures.

Maximum Contact Temperature

The European standard EN 50563 [R025] requires that the temperature at any touchable point of the enclosure surface be below the burn threshold. This threshold depends on the contact time; for eight hours and on any surface—untreated/finished metal, plastic or even wood—the standard is set at 43°C. For one minute of contact, the acceptable temperature ranges from 51°C for metal surfaces to 60°C for plastic and wood surfaces. For short contacts of one to ten seconds, a common value of 48°C is allowed for all surfaces. Now, as usual, you should add a margin to this value. My suggestion is not to have the case temperature anywhere higher than 40°C.

Questionable Published Heatsink Resistance Values

The values given by the manufacturers for the thermal resistance of a heatsink are usually far from the results you will get, even with the best configuration. The value you will find in a datasheet is commonly the thermal resistance that results with the heatsink when a channel ("duct") is used [R019]. This creates a chimney as shown in Figure 110. The chimney effect results in a heatsink resistance that is at least 20% lower than without the duct [R019]. Since customers select heatsinks primarily based on the ratio of thermal resistance to volume, manufacturers like to quote these lower "ducted" values with the chimney, but often do not note it.

In reality, it is not just the channel that is missing. Often the placement of the ventilation holes and the position of the heatsink itself is suboptimal. A good case is shown in Figure 110, where the heatsink is

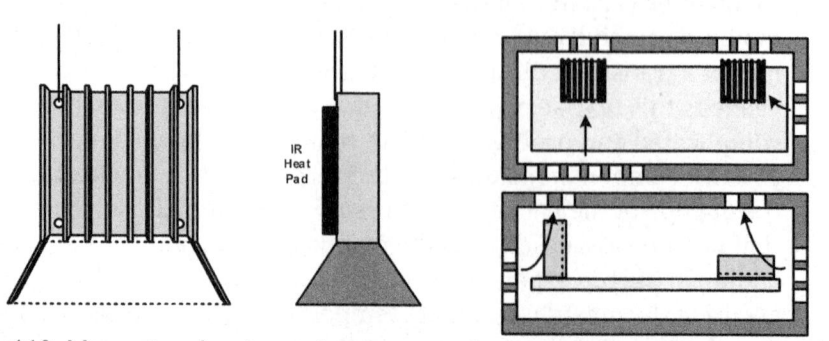

Figure 110. Measuring the thermal resistance of a heat sink in a chimney (left). Positioning and orienting heatsinks in natural convection (right).

placed on top of the vertically oriented PCB, with the fins also oriented vertically. This allows the hot air to leave the case immediately, creating an optimal natural convection path. Obstacles, such as large electrolytic capacitors, are usually encountered in this process. However, these can actually improve the situation by creating a turbulent airflow. A turbulent airflow dissipates heat better than a laminar airflow because it reaches every nook and cranny.

In practice, however, you may need to move the ventilation holes from the bottom to the side, as shown in Figure 110. This will render part of the heat sink ineffective. This loss of effectiveness is even more pronounced when the PCB is mounted horizontally; thermal resistance increases by about 20%-30%. In general, in my experience, you can get a conservative estimate of the real, final heat sink thermal resistance value by doubling the catalog value.

At sea level, a heatsink dissipates 70% of the heat by convection and 30% by radiation [R164]. At an altitude of 3000 m above sea level, the convection value drops to about 50% [R166]. At an altitude of 21 km, about 70–90% of the heat is removed by radiation, and the possibility of convection becomes correspondingly less significant.

Fin spacing is important for natural convection [R168] and should be greater than 6 mm [R164]. For forced airflow, fin spacing can be increased, but this is ultimately a trade-off for pressure drop. Fin waviness is insignificant for natural convection, but useful for forced convection, reducing thermal resistance by 20%–30% [R164]. Thus, there are 1.2–1.5 mm thin fins for natural convection and about 1.5–2.5 mm thicker fins for forced convection [R164].

Heat Sink Assembly in Production

An important aspect in the selection of a heat sink is the mounting method, especially if it is suitable for mass production. For a long time, the most common assembly was as follows: a screw, a washer, an insulating washer, the transistor, thermal grease, a mica plate, thermal grease (again), the heatsink, a thrust washer, and a hex nut. After all of this is assembled, the assembler must apply a defined screw force. If this is too high, the transistor body will lift off like a lever, with disastrous results in terms of effective thermal resistance. The die may even break. If the screw force is too low, the desired thermal resistance will not be achieved. Precise tightening force is even more important when a thermal pad is intended to eliminate the use of thermal paste. It needs to be quite high, but too high will cut the thermal pad, especially if there are burrs on the heatsink hole. Finally, the assembler must put everything together before soldering so that the solder joints are not subjected to too much mechanical stress. This requires a manual or selective (robotic) soldering step. If the wires need to be bent, this creates another potential source of error if the bending is not done correctly. This is a very complicated process for mass production. It is no wonder that premature failure of power semiconductors is mainly due to faulty heat sink mounting [R161]. Riveting would speed up this process considerably, but it presents the same problems discussed above.

Fortunately, there is a better way: You can use a clip mounting system with a TO-220 and similar THT components. Because the clip applies the force exactly where it is needed—above the die—the thermal resistance is equal to or less than the screwed solution [R161, R058]. The recommended force is between 20 and 50 N [R162]. Alternatively, an SMT package with an enlarged landing area can be used, as described below. Semiconductors in TO-220 and similar packages can also be clipped directly to the PCB if a heat sink is not required. The pins of these ICs are not designed to hold the component mechanically and may break under vibration or shock.

Heat Sink and Electrical Insulation

Imagine a buck converter where the diode is electrically connected to a heat sink. Conventional power diodes have a cathode on the heat sink. In a buck converter, the cathode is connected directly to the switching transistor. The heat sink then propagates all the switching noise. To a small degree, this would also happen if you chose a diode whose anode was designed to be mounted on a heat sink.

Alternatively, use a transistor with a grounded tab. However, when heat sinks are required, the currents are usually high, and ground bounce can easily occur.

Therefore, the usual rule is to isolate the heatsink from the hot device and leave it floating [R012]. Connecting the heatsink to ground would again introduce the risk of ground bounce EMI. It also creates an undesirable capacitive path for medium to high frequency currents from the hot device to ground. If the heatsink is kept floating, there is a second capacitance from the heatsink to ground in addition to the capacitance from the hot device to the heatsink. In series, the resulting capacitance will be less than if the heatsink were grounded.

Electrical isolation of the semiconductor from the heat sink usually degrades heat transfer. There are situations where the EMI aspect can be neglected and a transistor can be electrically connected to the heatsink. However, a bare connection should not be used; the resulting thermal resistance depends on the surface roughness, a parameter that is difficult to control in volume production. Instead, a piece of Grafoil™ can be used, as discussed in the next section. Although Grafoil™ is conductive, the heatsink can still be electrically isolated from the transistor if it is anodized. Black anodizing provides better heat dissipation and an electrically non-conductive surface. Scratches made during prototyping, such as when screwing a transistor's tab onto a heatsink, can inadvertently cause the heatsink to be electrically connected to the transistor. In contrast, for a defined electrical contact, it is necessary to remove the anodization at the location of the transistor, unless the heatsink has some provision for being electrically connected. However, as mentioned earlier, in most cases it is still advisable to leave the heatsink at floating potential.

Thermal Connection of the Hot Part to the Heat Sink

The thermal resistance of the various methods of connecting the hot component surface to the heat sink—dry, with thermal grease, or with an insulating pad—varies in the range of $R_{\theta CS}$ = 0.1°C/W to over 4°C/W, and is also highly dependent on the contact pressure [R161]. Before discussing this aspect, we should consider the following: For example, if we use a 10°C/W heat sink and improve the thermal connection of the hot part from 1°C/W to 0.5°C/W, the overall result will only change to 10.5/11 × 100% = 95.4% of the original heating. In most cases, such a marginal improvement is not worth the effort.

Due to the lack of alternatives, thermal grease has long been mandatory for low $R_{\theta CS}$ values of about 0.2°C/W without a mica plate and about 0.4°C/W with a mica plate. However, the correct application of thermal grease is an art. During vibration testing, portions of over-applied grease can escape from the joint into the PCB. There they bind dust and create stray current paths. Removing the excess paste requires alcohol or acetone, both of which should not come into contact with the plastic packaging of semiconductors because they can migrate into the package. Therefore, very little grease, but not too little, must be used to compensate for the surface roughness. Usually, the best amount must be found through a series of experiments, and the final result depends on the skill of the person applying the paste [R161]. This is clearly a process to be avoided in a quality-controlled scenario. Thermal grease is now only recommended for situations where contact pressure is negligible, such as connecting the processor chip to a heat pipe on a laptop.

Whenever possible, a defined solid thermal bond should be used. A 0.13 mm thin piece of Grafoil™ will give an $R_{\theta CS}$ of 0.1°C/W quite independent of contact pressure. However, Grafoil™ does conduct. Insulating silicone pads have values in the range of 0.3°C/W to 0.8°C/W (i.e., some are as good as the earlier mica-grease solution), but they require considerable contact pressure combined with the above-mentioned risk of cut-through.

How to Avoid Heat Sinks

Instead of having one field-effect transistor (FET) connected to a large heat sink, you can spread the load over several transistors and perhaps eliminate the need for the entire heat sink. This is possible with FETs because their channel-on resistance $R_{DS(on)}$ increases with temperature. Two FETs of the same type connected in parallel will automatically balance the applied load. A resistor of 4.7 Ω should be connected in series with each gate to prevent interactions between the transistors, known as gate ringing [R163]. Two FETs in parallel result in higher dynamic

184   Thermal Management

losses due to the doubled gate capacitance, but for power transistors the losses in the channel are usually orders of magnitude higher, and these are halved when using two transistors instead of one.

Pads as Heat Sinks

So far, we have focused on medium to high heat generation. For lower values, as discussed at the end of this section, an SMD device and an extended pad can be used. Figure 111 shows that by using a 70 μm thick copper pad, the thermal resistance can be reduced by more than a factor of two to a value of 30°C/W, albeit with a relatively large landing area. At 30°C/W and a junction temperature of 150°C, approximately 2 W can be dissipated at an ambient temperature of 70°C. An additional increase in pad size does not significantly reduce the thermal resistance. Thus, in the above case, the effective heated area is approximately (3 cm)². For a 35 μm copper thickness, it is about (2.5 cm)² [R012].

Of course, the transistor or IC must be in a suitable available package. D²PAK seems to be the best option for this [R015, R017, R027, R028]. Note that 70 μm copper is not normally available for pooled PCB orders.

Use of Thermal Vias

A landing pad that is much larger than the IC itself is often not feasible. What about dissipating the heat to the ground plane or to a heat sink on the other side of the board? Vias placed for this purpose are "thermal" or "heat" vias, with a recommended pitch of 1 mm [R012]. As simple as the idea sounds, it is not easy to implement in practice.

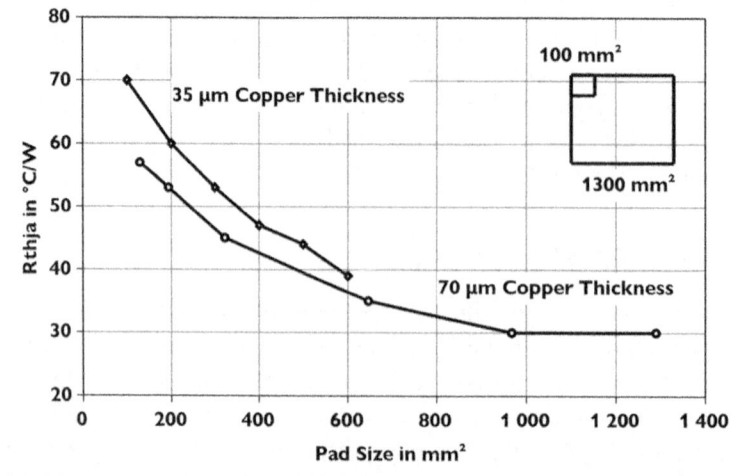

Figure 111. Using an extended landing pad as a heat sink for a transistor.

Placing vias directly under the thermal pad compromises the soldering process because the vias also dissipate soldering heat. It is usually not possible to adjust the thermal soldering profile. One solution is to move the vias to the edge of the thermal pad. These are still vias-in-pads. If their inside diameter is too large, the solder can still be sucked in and even seep through. It is recommended to choose a via size of 0.3 mm or less to avoid this [R012]. A small hole size also increases the chance of filled vias. One of my PCB suppliers specifies a negative tolerance of 0.3 mm for the finished via size. In pool production, the smallest available hole size is specified at 0.15 mm, with a wall thickness of typically about 25 µm. With this supplier, if I choose the minimum via hole size, I end up with a via that is at least about 50% filled and at best 100% filled. However, this introduces considerable uncertainty into the manufacturing process. In addition, the thermal conductivity of the copper alloy used for wall plating is usually not specified. Small changes in alloy composition can result in large changes in thermal conductivity. The usual remedy is to use many more thermal vias than the minimum required.

Alternatively, the vias can be covered with a solder mask. Both single-sided and double-sided masks can be used; the latter also fills the vias with solder mask but does not improve heat transfer. Via masking is a standard process available in the pool, so it adds only about 10% to the cost of PCB production. Note, however, that potential test points are lost when the via is masked.

Today, PCB manufacturers offer the option of capped, resin-filled vias-in-pads. Although intended for very dense designs, these vias can also be used for thermal purposes. When using such vias, the profile of the soldering process may need to be adjusted to accommodate heat loss. As technically attractive as this may be, it can quickly increase the price of the board by a factor of five. However, thermal vias to a ground plane on the opposite side are worthwhile; you can expect a 50–70% reduction in thermal resistance [R012]!

Use of Fans and Filter Mats

Natural unrestricted convection results in an airflow of approximately 0.3 m/s [R164]. Fans can be used to achieve ten times or more the air velocity. The volumetric airflow of a fan is either expressed in cubic feet per minute (CFM) or measured in cubic meters per hour (1 CFM = 1.7 m³/h). If you have an enclosure in which P_D watts of power are being applied and a rise in $\Delta\vartheta$ above ambient temperature is allowed, the theoretical minimum airflow required can be calculated by Formula 4.

Let us take the example of a typical axial fan with a cross-sectional area of 80 × 80 mm² and a typical airflow of 69 m³/h, such as the Ebmpapst 8412N. This allows us to keep the inside of a cube with a side length of 80 mm and 100 W of internally generated heat at only +4°C above ambient temperature. However, Formula 4 does not include any derating. In my experience, the theoretical number needs to

$$airflow = \frac{P_D}{\rho \cdot c \cdot \Delta\vartheta} \frac{m^3}{h}$$

Formula 4. Calculation of the theoretical minimum required airflow based on power dissipation P_D inside the enclosure, air density $\rho \approx 1.3$ kg/m³ at 30°C and standard air pressure, heat capacity of air $c \approx 1000$ J/(kg°C) at 30°C, allowable internal temperature rise $\Delta\vartheta$ above ambient temperature.

be doubled. In our example, this results in an expected temperature rise of +8°C.

Try to create a top-down forced airflow, as used in tower PCs, to prevent dust and lint from being drawn in. If this is not possible and the forced airflow is from the bottom up, a filter mat must be used. A coarse dust filter that retains particles larger than 10 µm—hair, sand, insects, textile fibers—is sufficient. Filters cause pressure drop. For the coarse filter mentioned above, the pressure drop at 1 m/s air velocity is about 30 Pa when a fabric with a filtration efficiency of 80–90% is selected. The pressure drop is linear with air velocity. This straight line can be transferred to the fan curve to determine its operating point, as shown in Figure 112. Finally, the air outlet should have at least the same opening area as the inlet.

I was involved in a project where six 80 × 80 mm² fans were mounted under several vertically placed large circuit boards. So the airflow came from the side of the ICs. Except for some electrolytic capacitors, all the elements were flat and low. After assembly, we measured the efficiency of this airflow cooling to be only 83%. This is why

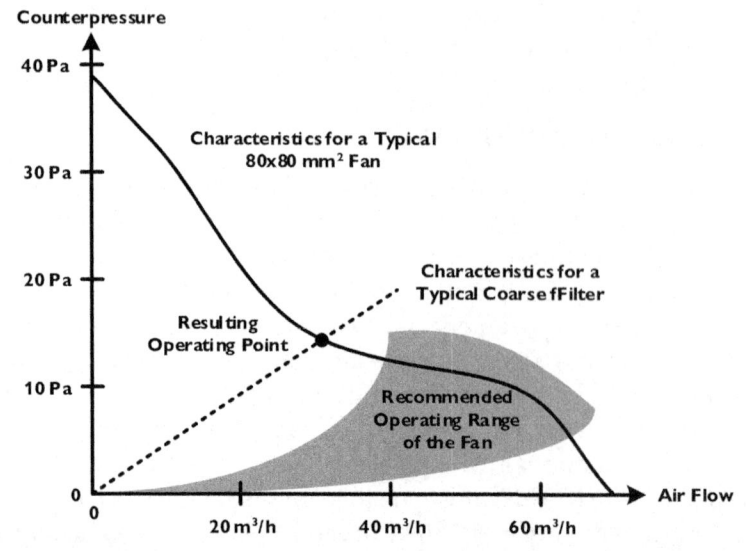

Figure 112. The operating point that results from combining a particular fan with a particular filter.

motherboard manufacturers mount fans directly on the processor and do not blow from one side. With an airflow directed at the top of an IC, heating can be reduced to at least 80% and at best 50%.

Cooling Systems, Condensation, Peltier Elements

When passive methods, including those using fans, are not sufficient, a cooling system with or without a compressor must be used. These systems are usually large, noisy, have moving parts, and must be placed in a specific orientation for operation. Active cooling provokes dew formation if the cooling rate is greater than -1°C/min or 6% relative humidity per hour, or if you cool below the dew point. Therefore, the cooling rate must be controlled.

From time to time, someone brings up the idea of using a Peltier element, such as those found in beer coolers. However, the disadvantages of such a solution usually lead to a no-go. First, the heat does not disappear; in fact, the Peltier element increases the heat by a factor of about 2.5 due to its poor efficiency. To dissipate 10 W of heat, the Peltier element needs an additional 15 W. In the end, you have to get rid of 25 W of heat on the cold side of the element. If this cold side can be touched by a user, the temperature should not exceed 40°C (see above). This will most likely result in a huge heat sink.

What if things are upside down? The Peltier element is powered by a voltage source; instead, a resistor can simply be connected to its terminals. This phenomenon, called the "Seebeck effect," really works. A temperature difference between the hot and cold sides of a thermoelectric cooler creates a voltage across its terminals. A connected resistor carries a current, allowing the heat to be electrically dissipated to where it is least disruptive. Unfortunately, the efficiency of this method is about 3% to 8%, which is even worse than the Peltier effect.

Creating a Thermal Prototype Early

The many ways of dealing with heat in semiconductors discussed so far, and their sometimes imprecisely calculable results, make it necessary to do experiments as soon as possible. For a thermal prototype it is advisable to buy a lot of different heat sinks. This is not the time to be frugal; a collection of heatsinks will also be useful in later projects.

Knowing how thermal conductivity scales with changes in size can be helpful in making a selection. For extruded heatsinks, the thermal conductivity increases only with the square root of the extension of the fins in the extrusion direction [R018]. The increase in fin height follows the same pattern. In both cases, you get half the thermal resistance with about four times the volume of the original heatsink [R018]. Increasing the number of fins improves the thermal conductivity linearly for the first few additional fins. However, as the heatsink becomes wider, the ends of the heatsink move further and further away from the heat source. As a result, they no longer contribute to a reduced thermal resistance value, until they make no contribution at all.

In terms of shape, extruded heatsinks are not the best solution. They tend to produce laminar airflow. However, turbulent airflow is much better at removing heat from even the most remote corners [R164]. Therefore, pin-fin heatsinks are more efficient than other extruded heatsinks of the same size. Unfortunately, the former are also the most expensive because we cannot make them by extrusion. Folded fin heatsinks are an alternative; their price is between that of pin heatsinks and extruded heatsinks, and they can still have ten times lower thermal resistance than extruded heatsinks [R164].

After ordering a few heatsinks, the focus should be on a preliminary enclosure. If the designers are overwhelmed and do not have the time, an electrical engineer can now design it himself. It took me no more than two hours to learn how to draw a very simple rectangular enclosure with circular openings for 3D printing. Of course, all the details are missing, such as snaps for mounting the PCB or domes. However, these subtleties have no significant effect on the thermal behavior; in fact, they are a hindrance to experimentation.

Now take a single-sided bare board and drill four holes near the corners to mount spacers. Using spacers instead of a designed part on the case to hold the PCB allows you to change its position inside the case. The copper side simulates the ground plane. If the hot element is a transistor with a heatsink mounting tab, drill a hole where the heatsink will be placed. Position the heat sink and screw the transistor onto the heat sink with a thermal pad in between. Connect the transistor to the power supply with the cable on the outside. The thermal model is almost finished. It only needs to be equipped with the sensing elements as described in the next section.

Even if you are experimenting with SMD transistors and extended landing pads, you do not need a fabricated PCB. A cutting tool or hand router can be used to create the desired pad from the copper layer and solder the SMD to it. Alternatively, a piece of copper foil of the appropriate thickness can be taken, cut to size, glued to the unlaminated side of the PCB, and soldered to the SMD. It is also possible to simulate thermal vias by drilling holes and passing thin wires through them. All of this gives only a rough estimate, but this is often valuable, especially if you have underestimated the thermal situation.

Temperature Measurements

The first thing to do is feel and smell. If you cannot put your finger on an IC, it is probably hotter than 50°C. If you can smell the characteristic smell, the chip is approaching 70°C [R166]. However, the goal of temperature measurements is to determine the temperature of the die, since that is the limiting given value. Unfortunately, this is not so sim-

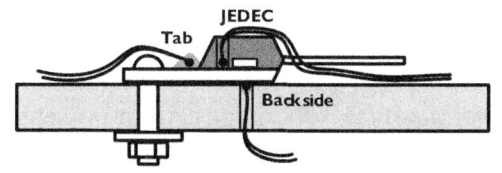

Figure 113. Three different ways to approximate the die temperature for a transistor in the TO-220 package.

ple. Figure 113 shows three different locations for measuring the temperature of the transistor in the TO-220 package when using a thermocouple. According to JEDEC, one location requires drilling a hole in the package [R165]. If the transistor package temperature is measured at the tab, the difference from the JEDEC position reading can be as much as 1°C/W [R161]. Alternatively, a hole can be drilled in the heat sink and the backside temperature measured. This will be higher than the JEDEC value, indicating a more conservative approach. Adhesives with good thermal conductivity should always be used. Do not just use tape to hold the thermocouple in place. The entire weld area must be covered by the adhesive to avoid averaging the surface and ambient air temperatures. However, the adhesive plug should not be larger than 2 x 2 mm² or you will again have too much ambient surface area [R160]. Use a wire smaller than 36 AWG or the thermocouple will affect the reading due to heat dissipation. Run the wire along the hot surface to the edge of the PCB and secure it with tape if necessary. Do not remove the thermocouple wire from the PCB until it is at least 25 mm from the hot spot. Improper thermocouple setup can easily result in errors of up to 50%. At a minimum, perform measurements with two different heatsinks and two different contact pressures. If the junction-to-case value is relatively constant, it is reliable.

When measuring with a handheld multimeter, it does not matter if the adhesive is also conductive (i.e., it could be soldered to the board under test). However, when working with a data logger or other device connected to the mains, a non-conductive adhesive must be used to avoid ground faults.

For a large heat sink, thermocouples with a copper washer on the end can be considered. They have high temperature resistant Kapton tapes for quick mounting. However, they should not be used to measure a single temperature point, such as the top-center temperature of a microcontroller or FPGA, or for the TO-220 tab temperature measurement mentioned above. The washer will absorb too much heat and give you an overly optimistic result.

In addition, the die temperature can be measured directly by the voltage drop across an ESD protection diode if present on an unused pin or on a pin at constant potential using a forced constant input current [R156, R157]. The chip select or reset line can be used with a current source of approximately 100 µA below 0 V or above the supply voltage [R166]. However, if the ESD diode characteristic is unknown, which is usually the case, it must first be measured with an unpowered IC. Then it must be verified by another set of measurement points that powering the IC does not change the curve obtained. In addition, the temperature dependence of the forward voltage cannot be determined from the diode characteristics. The slope of the dependence is often about -1.7 mV/°C, but for a given device it may be -1 mV/°C or -2 mV/°C. This means that further measurements are required when the unpowered IC is placed in an oven. Due to the linearity of the slope, it is sufficient to take two measurements in addition to the room temperature (e.g., at 60°C and 120°C). This does not take too long because the oven heats up quickly and the IC follows its temperature almost instantaneously. This can be more of a problem if an oven is available at all, just at that moment. Only then can the die temperature be measured using the forward current of the ESD diode.

The method can also be used as an internal overtemperature sensor if not already implemented in the IC. For this purpose, the voltage-temperature relationship is recorded for 10 or more samples and the mean curve and the maximum standard deviation are calculated. The latter indicates the expected accuracy of the temperature measurement over the entire batch. If the standard deviation is 3% or less of the corresponding stress temperature value, the averaged curve is usually considered accurate enough for the entire lot [R158]. If this is not the case, individual voltage-temperature curves should be used or the method should be avoided. The required low-current, temperature-independent precision current source can be implemented with an adjustable shunt voltage reference or with a precision voltage reference and an operational amplifier [R159]. Because the ESD diode is at the edge, the chip may be hotter in the center. It is preferable to use a margin of about 25°C over the maximum die temperature [R166].

With an open housing, an infrared (IR) gun with a maximum spot diameter of 4 mm can be used. Laser IR guns are usually unsuitable because they are designed for longer distance measurements and their spot sizes are too large [R023]. The IR gun should be adjusted to use the appropriate emissivity value. This is only possible with more expensive guns; inexpensive guns are set to a value of 0.95. The latter is possible because all materials except uncoated metal surfaces—oxidized or not—have emissivity values in the range of ±0.5 to about 0.90, regardless of their color [R023]. Obviously, correction will result in a more accurate value. If the exact emissivity of a surface is not known, consider painting it with a coating of known emissivity. The most accurate results in terms of surface temperature are obtained by making an IR measurement with the correct emissivity value [R160]. Of course, the hot spot must be accessible.

Note that it can take more than an hour for a large heat sink to reach thermoelectric equilibrium. If you rush, you may misjudge the severity of the situation.

The easiest way to determine the case's internal temperature is to use irreversible temperature markers, also called temperature-sensitive/indicating strips. The "watch-style" rings are the smallest, with a diameter of 14 mm. These are not sold by the usual suppliers of electronic components; they must be obtained from other suppliers. Be sure to buy irreversible ones unless you have access to the inside of the running device's case.

Thermal Simulations

Thermal simulation of electronic circuits is about finding the true maximum temperature value for each component. Manufacturers of thermal simulation software advertise that their tools eliminate the need to create a thermal model. This is true when they describe their simulators as very powerful. They need to be very powerful because, due to widely varying component heights, there is usually turbulent airflow, and if this is not modeled correctly, the results will be completely wrong. Such calculations sometimes require so much computing power that manufacturers offer access to their server bank for uploading a problem to be solved. So, the available computing power is no longer a problem, but there are other issues, namely the reliability of the results. The problem lies in the model parameters [R029]. They usually require more time to properly define the problem for simulation than to produce a simple thermal prototype. Despite the best problem definition, simulators still fail to find the exact value for real situations [R029]. The simulated results must be verified with a real test setup, which of course begs the question of why you should do a simulation at all. To my credit, simulations generally show trends correctly. Starting with a verified model, simulations can quickly evaluate the effects of changes to the model.

After all, thermal simulators are not suitable in the very areas where we could use them most to evaluate the very time-consuming effects of thermal cycling on an electronic system.

6 Testing and Verification

This chapter deals with tests that you perform on prototypes or that are performed according to your instructions on prototypes, preproduction, or production.

Thinking About Testability from the Start

On one project I was involved in, five of us engineers spent three weeks trying to figure out why the first prototype occasionally failed. In the end, it was a temperature-dependent timing problem. During the time we spent trying to find the problem, we had to modify the board several times just to run certain tests. We learned that it is best to allow maximum testability for the first prototype. For every new element you add to a circuit, think immediately about whether and how it can be tested. Add test capabilities to the schematic and layout as you go. Do not think about adding them later. There may not be time, and there may not be space on the board. Effort in the design phase is more acceptable to customers and supervisors than effort in debugging. However, do not talk about these measures unnecessarily. The customer and the boss sometimes expect the very first prototype to work perfectly. There are all these simulation tools, there are the design rules, and what can go wrong?

Do not add components or structures that you cannot test. This may sound strange, of course you would not. But consider this common case: a PCB antenna is to be used for Bluetooth connections. To properly test this antenna, you need a network analyzer, an expensive piece of equipment that is usually only available to companies that design high-frequency circuits on a regular basis. You also need someone who knows how to make the critical measurements and who can tell you how to improve the antenna design. Consider hiring an RF design service provider if your company does not have expertise in this area.

If you must use equipment other than that available within your design team, reserve or allocate this equipment as early as possible. Make sure a climate chamber or vibration table is available as soon as you know you will need it. Do not wait until the prototype is populated and basically working. It is easier to move a registered assignment back in time than it is to get a new assignment as soon as possible. This means that to make a meaningful reservation, a rough test plan must be sketched in parallel during the schematic phase. No one likes to do this at this or any other time when you are in the most interesting, creative phase of development, but it is a typical case where an omission will cause problems later.

Enabling maximum testability can be a challenge in a dense design. One way is to make the first prototype larger than the target size. Finally, geometric optimization is usually a step whose duration can be well estimated. On the other hand, debugging has no time limit, and experience shows that the more compact the layout, the longer it takes. However, if you oversize the first prototype board, you lose the opportunity to check the interaction with the mechanics early on. It also raises doubts about whether the target size can be achieved at all. One solution is based on the experience that the first prototype of a new device is never exactly the same as the final PCB. If there is an additional PCB run anyway, we can incorporate critical components and their test measures on the first board and use breakouts inside or even outside the device for non-critical parts. With a working first prototype, we can remove all test fixtures not required for post-production testing and copy the breakout layout(s) to the PCB.

Basic Testability Measures

The following are some detailed aspects of maximum testability:

- All control and data signals and all supply voltages can be accessed from one side of the board via test points or connectors. Ideally, every electrical node is accessible via a test point or test plug. When troubleshooting, you will appreciate every bit of information.

- Functional units can be separated and studied separately.

- Instead of signals generated on the board, "artificial" signals can be applied; examples follow.

For digital systems, additionally:

- Internal interfaces (I^2C, SPI) are accessible via connectors.

- There is a status indicator, usually one or more LEDs. The RHOM SML-P11x series has only a 0402 size and 1 mA current consumption. In addition to turning it on and off, you can also flash the LED with different rhythms to indicate the different states the program is in. This will be sufficient in most cases. One idea is to have a blinking routine for the idle state. If the LED stops blinking, the firmware is probably stuck in an interrupt routine.

- There is an option to activate a process (e.g. a push button). The KMR 2 series from C&R is no longer than a 1206 resistor and only about 1 mm wide, so it fits almost anywhere. Of course, you do not have to include such test buttons in the final product (i.e. their cost is not important). It is only the space that needs to be available if no PCB redesign is required for the final product.

- A UART that sends status messages uses only one pin of the microcontroller. It is such a simple debug interface, but it can be immensely helpful.

- There is a reset button.

Preparing for Production Testing

Consider from the beginning how the device will be tested after production, and it is best to talk to the assembler early on. The schematic may need to be modified:

- Do not connect chip enable, reset, and unused inputs and outputs directly to VCC or GND, but with a 100 kΩ pull-up resistor or a 100 Ω pull-down resistor, respectively. This allows the ICT to turn on individual ICs and test them individually.

- Connect unused pins of ICs to test points. Otherwise, a short between an unused pin and an electrical network cannot be found.

- Do not connect the reset of a microcontroller) directly to the watchdog, for example, but via a 100 Ω resistor.

- Route the clock module through a multiplexer so that the clock can be turned off.

- Low capacitance on stabilized control lines (e.g. debounce circuit) allows faster testing.

Adjustments to the layout may also be necessary and are discussed in the following section.

About Those Test Points…

We know from experience that test points smaller than 1 mm are difficult to probe; it is difficult to stay on them with the oscilloscope probe tip. It is better to use test points larger than 1 mm (e.g. 1.27 mm). Do not forget the test points for all supply voltages and some ground test points scattered around the board. The latter come in very handy when you are doing a measurement; you need access to the ground potential every time. Consider placing a ground test point about 7 mm away from an important signal test point, for using the small ground pin that

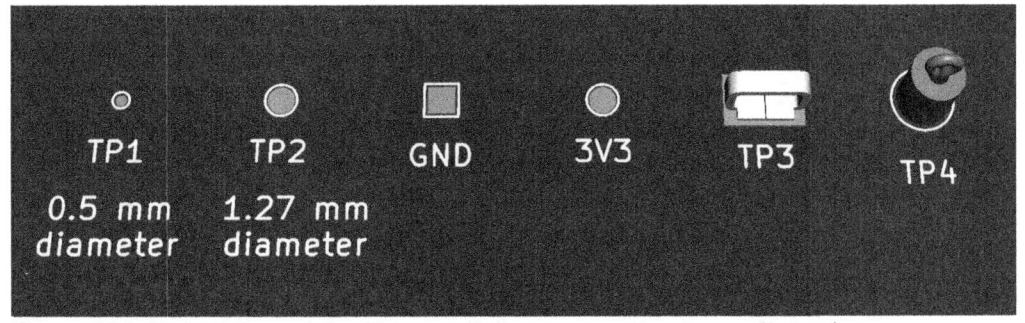

Figure 114. From left to right: Too small diameter test point, well sized test point, ground test point (GND), supply voltage test point (VCC), SMD test point, and THT test point.

comes with the oscilloscope. Otherwise, if you use the common ground clip and attach it somewhere far away from the test point, you will create a large ground loop, and you may introduce ringing into your measured signal. Then it looks bad when it is in pretty good shape with the short clip or no measurement. Figure 114 shows some of the mentioned test point variations.

If a bed-of-nails tester is to be used, you may want to check if it requires the test points to be on a particular grid (e.g., a 1.27 mm grid). There is nothing worse than reworking a dense layout to move all the test points to match a grid!

Bed-of-nail testers require at least two large tooling holes per board and multiple pressure points.

For test points where a DC voltage is measured with a multimeter and clamp probe, use only THT-mount test points, not SMD test points. They will tear off quickly. THT-mount test points take up more space on all layers, but they are robust, which is important in the sometimes chaotic development phase.

It is tempting to use uncovered vias as test points, since there are usually many on the board anyway. But consider this:

- To probe vias by hand, they must be larger than usual. As mentioned above, the diameter should be larger than 1 mm. This is especially true when the oscilloscope probe tip is larger than the via hole size. Check that you can sit comfortably on the via. If you can easily slip off the via, you need to increase its size. It is usually not possible to increase the size of all vias, only some of which will be used as test points. One workaround is to replace the vias to be used as test points with a special element that has the shape of a via but is larger in diameter.

- There are mixed experiences using vias with bed-of-nail testers. I have met engineers who have never had a problem with this approach; others have stayed away from it, reporting that the needles get stuck in the vias. Either way, check the via and hole size requirements with your bed-of-nails tester. Special needle tips must be used for this probing. Be sure to note which are via test points and which are normal test points.

- As suggested above, replace the ordinary via to be used as a test point with an explicit via test point element. This will make the via test point appear in the schematic. Otherwise, the test point may be inadvertently lost if the layout is reworked and the via is no longer needed. Also, each test point should be clearly labeled in the schematic, which is difficult to do with a simple via.

Tiny Holes to Save You Embarrassment

If there is space left on the PCB after the layout is finished, sprinkle some 0.3 mm holes on it. A PVDF-coated AWG 30 wire will slip neatly through such holes. If you need to route a signal from one side of the board to the other, you do not need to wrap the wire around the edge, creating a visible, somewhat embarrassing patch. Instead, the traditional yellow patch wire slips elegantly through an existing hole, which may even look like it was planned.

Ideas for Measuring Prototype Current

In general, current measurement is important for low-power circuits, e.g. to determine whether the microcontroller is in sleep mode. Supply currents also provide important information about the proper functioning of ICs. ICs that are defective or incorrectly wired in some way often have significantly higher current draws than specified in the data sheet. Two methods of measuring current are discussed below:

- "Bridge" variant: two pins of a pin header are placed in the current path, splitting it. The pins can be bridged with a jumper. Advantage: DC measurement is possible without conversion; disadvantage: the pins take up space, both horizontally and vertically, and the jumper can become loose (i.e. not suitable for vibration testing).

- Low-impedance precision shunt resistor. Advantage: cannot become loose; disadvantage: current must be converted to voltage. Noise limits the accuracy of this solution at low currents. High currents may require a physically large shunt. There is no one good value when measuring low and high currents.

In the supply design chapter, Table 18 lists the main IC types for current sensing with a shunt. Figure 37 illustrates layout variations for current sensing.

Collected Ideas for Improving Prototype Testability

Here are some specific ideas for improving testability. Figure 115 shows solutions for separating elements in a signal processing chain. In the first case shown, the connection of an oscillator to a mixer is separated in the layout of the first prototype, and each side is provided with high-frequency coaxial connectors (e.g., SMA types). Both units can be tested separately; if they work, a coaxial connection is made. If it works, the connectors and bridge are replaced with a PCB trace in the second layout pass.

The second case in Figure 115 shows the measure of separating units by placing a 0 Ω resistor in between. Note, however, that the 0 Ω resistors also have a maximum current specification. When separating a digital signal with a (removable) 0 Ω resistor, the third case in Figure 115 indicates that the digital level may need to be defined with a pull-up resistor or pull-down resistor.

Sometimes it is advantageous to be able to disconnect an entire data bus. In the case of a unidirectional bus, this is easily done with 245 logic ICs. In principle, the same function block can be used to operate a bidirectional bus, since the direction of the data flow can be switched. For example, if you place a connector between the microcontroller and the 245 IC, you can easily access the bus output with a logic analyzer and feed in artificial patterns to check the reception of data by the microcontroller.

With digital communication between ICs on the board, troubleshooting is greatly facilitated, if not made possible, by logic analyzers. It is recommended that you think about this when you are drawing the schematic. How can you test an SPI connection, an I²C bus, or a parallel data/address bus? Think about where to place a connector for a simple logic analyzer connection. To save space, build an adapter from the analyzer's 2.54 mm pitch connector to a fine-pitch connector with half the pitch (1.27 mm) and size, or an even smaller connector placed on the PCB. However, you should not use a logic analyzer test connector to collect signals with additional long PCB traces from all over the board, even if it is possible from a layout standpoint. Long test leads can pick up noise that would not otherwise occur. Use multiple small connectors instead.

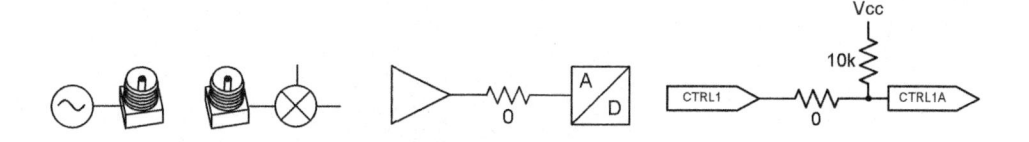

Figure 115. Left: Solution for separate analysis of an oscillator and a mixer by separating them and using an SMA connector. Middle: Solution for separating two functional units using a 0-Ohm resistor. Right: Solution to define a digital level when the 0 Ω resistor is removed.

For troubleshooting, but also for performance testing, it can be very helpful to be able to run a system that has its own clock source with a slower or faster clock. All you need is a way to feed in an external clock. This is best done with a mux/demux bus switch (e.g. 7SB3157) instead of a logic multiplexer (e.g. 7SZ157); the former has a smaller delay and practically no power consumption.

Prototype Microcontroller Programming

We have not discussed how to program the microcontroller during the development phase. Figure 116 shows some variations.

The most common method is to use a pin header, usually with 2.54 mm pin spacing. Even with only six pins, this is quite a large component compared to the rest of the board. If you go with a THT header, this may cost you too much space on all layers combined. However, I would advise against using an SMD header unless you are going to glue it to the board. In my experience, SMD headers are easily torn off the PCB. It is a large connector where the SMD pins are simply too small to be mechanically secure enough.

I have seen many boards using TE Connectivity's Micro-MaTch™ series of connectors. Because of the lower height, the torque generated when pulling on a connected cable is less than with a traditional header. In addition, the solder pad size is larger. It is also a polarized connector, which prevents accidental mating, which can easily happen in an often complex design environment. If you need it (e.g. for vibration testing), there is even a latching version available. The mating connector is an insulation displacement connector, so you can easily assemble it, albeit with a 1.27 mm ribbon cable. This means that you usually need an adapter if you want to connect it to the 2.54 mm connector of a logic analyzer.

I discourage the use of more compact connectors for programming, such as the Hirose DF12NC series. These connectors are perfect for

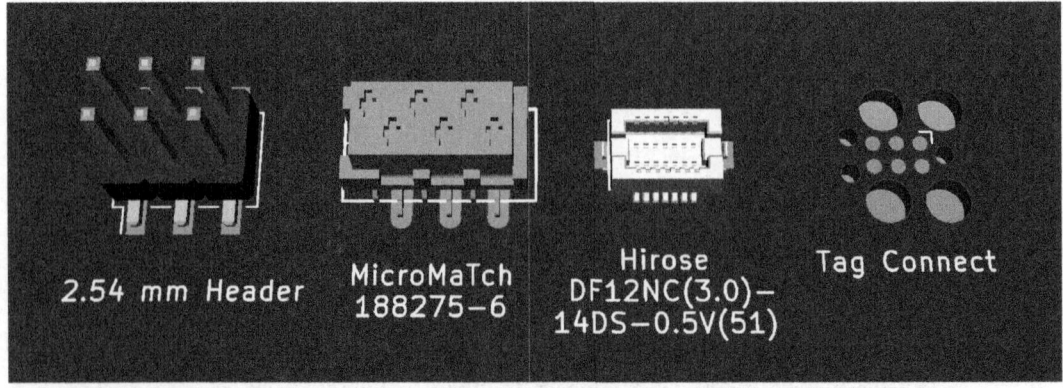

Figure 116. Connectors from left to right: 6-pin header, 6-pin Micro-MaTch™, Hirose DF12NC, Tag Connect™.

attaching a shield to a PCB, but they are not made for many disconnects. If you disconnect a wire, you can easily rip off the jack as well.

An interesting variation is the Tag Connect™ system. In its smallest form, it consists of six pads like test points and three orientation holes. You must press a probe with six bed-of-nails needles and three orientation needles against the board while programming the microcontroller. Most of the time this will not be feasible, especially if you reload the firmware often. If you need four more holes, there is a probe variant with a latching function. However, with these latching holes, the total required connector size on all layers is about the same as a six pin THT header. Clever people therefore created a latch that could be attached to the orientation pins. With this fix, you have the smallest possible programming connector again.

Desktop Temperature Checks

The chapter on thermal management deals with precise temperature measurements. Such measurements are usually unnecessary in the very early stages of development. It is often more important to know which elements or subcircuits react sensitively to temperature changes. This does not require a time-consuming setup in the oven or climatic chamber. Instead, an industrial fan and cooling spray can be used. Industrial fans are those in which air temperature and velocity are selectable, preferably digitally. Small nozzles allow heat to be applied to individual ICs. All analog circuits should be specifically targeted, and all circuits that provide reference voltages. Using a cooling spray is a haphazard method since the temperature cannot be controlled. It is usually much too low in the -50°C range. However, as the component warms up, you can watch the circuit return to normal operation. An IR gun can be used to check at what temperature this happens.

Thin traces and ICs in miniature packages such as the SC70 can get hot without you noticing. Therefore, it is a good idea to occasionally look at the board with a commercially available, inexpensive standard infrared camera to detect such problems. A more expensive high-resolution infrared camera will allow you to see the temperature rise of traces carrying milliampere currents. Needless to say, this can be very eye-opening. I know a fellow engineer who regularly debugs his boards using this method.

Your Own EMC Tests Part 1

EMC test centers are usually busy, which leads to two important recommendations:

- Make an appointment as soon as possible. Wait times can be up to three months. Don't wait until the first prototype is built and running before you book. Clarify the availability of the test center at the beginning of the project and book according to the expected date when you have a DUT available. It is usually easier to reschedule an existing appointment than to get a new one.

- By performing the following tests yourself before visiting an EMC test center, you won't have to wait until you get there to find trivial errors. Although such faults can often be quickly remedied in the field, e.g. with additional capacitors or ferrite beads, valuable time has been lost up to that point and at the EMC lab itself.

The most important test you should perform before visiting an EMC test center is the electrostatic discharge test. This is because it can potentially destroy the device under test. There is nothing worse than waiting weeks for an appointment at an EMC test center, only to have your board fail after the first discharge test! In addition, ESD testing can easily be done by yourself with an ESD gun, which is a handy and inexpensive tool. Have your company buy one if you design electronics for them. The pre-tests you do on the lab bench will save your company a lot of money in the end.

Therefore, it makes sense to discuss the 61000-4-2 ESD test in more detail here. First, the test level must be defined according to Table 31. Many standards, generics, and products require Level 3 (i.e., 6 kV contact discharge and 8 kV air discharge). In practice, however, most companies generally require Level 4, even if it is not required by the applicable standards [R038]. However, this is of no concern to the designer,

Test level	Contact discharge (needle tip)	Air discharge (thumb tip)
1	2 kV	2 kV
2	4 kV	4 kV
3*	6 kV	8 kV
4**	8 kV	15 kV

* Household, light industry. ** Test level selected in practice.

Table 31. Test levels according to IEC-61000-4-2 [R032].

since the circuitry required for Level 4 protection is no different from that required for Level 3 protection. Therefore, from the designer's point of view, it is advisable to test to the highest level in order to have a product that is as safe as possible.

The test procedure is shown in Table 32. Note that for air discharges, all lower levels must be passed before moving to a higher level. This is because air discharge devices may pass the highest test level but fail the lower levels. Such results are plausible when one considers that the flashover paths at a higher voltage may be different from those at a lower voltage. It has even been noted that the speed at which the tip of the thumb is brought to the device is important [R040]; the highest possible speed is recommended. Robots are also used for reproducibility of test results [R042].

In general, some discharge objectives are explicitly excluded [R040], as follows:

- Areas that are only accessible during service, even if the service is performed by the end user.

- Areas that are inaccessible after assembly or installation. This means that the connector of the box shown in Figure 54 in the Robust Interfaces chapter should not be inspected, but it was inspected anyway.

- Pins on connectors with metal frames. Without a metal frame, air discharge must be attempted on the pins.

- Connectors that are labeled as ESD sensitive because they cannot be protected from ESD for functional reasons (e.g. antenna connectors).

However, product standards may override these exceptions. For example, automotive products must test each connector pin individually, even if the connector has a metal frame [R042]. Even if no applicable standard requires it, you may want to remove these exceptions yourself to verify that you have designed a robust device. For example, when a repair is performed, an ESD-safe environment may not be automatically guaranteed, and you will have failures after a repair, which is especially annoying to customers.

In the case of the plastic housing, an air discharge is attempted on the housing, but may not be achieved, in which case the horizontal and vertical coupling plate tests [R042] must be used.

Table 21 in the Robust Interfaces chapter lists the minimum pulse numbers. Product standards may require much more. For example, the EN 55024 CISPR24 standard for information technology equipment requires 200 positive and 200 negative pulses at a minimum of four selected points. I recommend that you pre-test your ESD gun on the circuit with more than the minimum number of pulses; for example, I

Contact discharge on conductive elements	Air discharge on insulating elements
Metal connector frame. Metal switches. Metal display bezels. Joints for metal enclosures. Coupling plate* 0.5 × 0.5 m².	Pins of connectors without metal frames. LEDs. Keypads. Openings in the case. Randomly selected points on the case.
Minimum 10 positive discharges, 10 negative discharges Maximum pause of 3 seconds between discharges.	
Maximum test level only.	All lower test levels must also be tested and passed first.

* At a distance of 10 cm from the unit, once vertically and once horizontally.

Table 32. Test procedure according to IEC-61000-4-2 [R032].

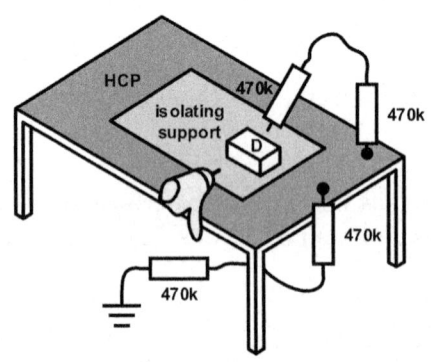

Figure 117. ESD test with a plastic case using a horizontal coupling plane (HCP).

Criterion	Description
A	Unit is operating continuously as specified.
B	Temporary loss of function, self-healing, no corrupt data transmission, or faulty data storage.
C	Temporary loss of function, operator intervention required.
D	Permanent loss of function.

Table 33. Result categories according to IEC-61000-4-2 [R032].

scan the device for sensitive points up to a rating of 20 pulses per second. However, this cannot be done on an isolated device, as explained in the next paragraph.

For a battery-powered device or an isolated device with no ground connection, the stored charge usually does not dissipate fast enough between two pulse applications. The device will still be at a high voltage level (e.g., 4 kV) when the next test pulse occurs:

- A subsequent discharge of the same polarity is weaker because the differential voltage between the generator and the equipment is smaller, e.g. 8 kV–4 kV = 4 kV.

- A subsequent discharge of reverse polarity is stronger than defined because the differential voltage is higher,
e.g. 8 kV–(-4 kV) = 12 kV.

The preferred solution is to connect cables with a 470 kΩ resistor at each end to the horizontal coupling surface on one side and touch the discharge point with the other end after each ESD application (see Figure 117) [R041].

Do not use a cable without resistors, otherwise a hard discharge will occur, which acts as a new test charge. Alternatively, you can brush the device with a carbon fiber brush grounded through a 470 kΩ resistor after each ESD. Finally, you can simply wait until the device has discharged, which can be checked with a galvanometer.

The results of an ESD test according to IEC 61000-4-2 are evaluated according to the four possible criteria in Table 33. It all depends on the type of product that needs to be fulfilled. In general, the following can be said:

- If the device controls a machine, or if there is a hazard to people if the device malfunctions, Criterion A applies.

- The industry standard is Criterion B [R041].

- Criterion D is equivalent to "test failed".

According to Table 33, Criterion A means that the device continues to operate without interruption. No reset may occur and no control signal may change. However, this can happen, for example, at a circuit node with an impedance level that is too high. Designers typically keep the node impedance below 1 MΩ, sometimes even below 100 kΩ. This is the reason not to use high impedance pull-up resistors, even in battery powered devices. Consider placing a capacitor at nodes that carry DC-like signals. For example, a capacitor of about 100 nF is a must on each reset input pin.

Your Own EMC Tests Part 2

There are EMC tests you can perform that are so simple and effective that there is no excuse for not doing them:

- A chatter relay can be used to implement a jammer with a bandwidth greater than 1 GHz. For example, a DC relay with a momentary on function is used and connected to a DC voltage as shown in Figure 118. A wire carrying a signal or power to the DUT is placed along the relay supply line to capture the spurious. Chatter relays are even used in certain standards, such as J1113-12 (automotive EMC) or DO-160 (aircraft). A bipolar TVS diode can be used to limit the level of voltage spikes on the relay supply.

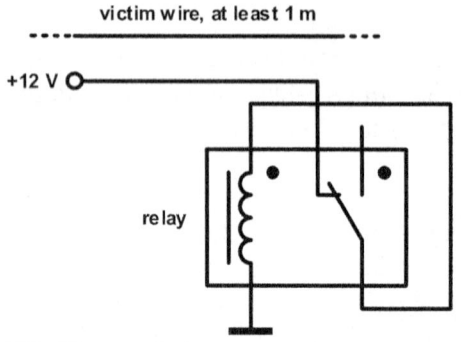

Figure 118. Chatter Relay.

- A 3 VDC motor connected to two AA batteries via a 1 m cable has an interference spectrum up to 750 MHz. As with the chatter relay, place the victim wire along the motor supply cable.

In principle, some EMC measurements are also easy to make: a wire connected to the oscilloscope input is already a probe for the electric field; if you short it to ground, you can measure magnetic fields. Probes of different sizes are easy to make and can be insulated with heat shrink tubing so you can get very close to the board. The difficulty with these measurements is interpreting the results. It is unclear what is acceptable and what is not.

PCB Markings

Preferably, from the first prototype, you should provide space for each PCB:

- PCB version

- Assembly variant

- Serial number

- Check boxes

- Project name, date and company logo

This is best done early, as these elements also need space. Non-critical components and functional groups can be placed accordingly.

Pre-Production Series Lot Size

For the first or initial prototypes, only random tests are possible. Their main purpose is to identify a major problem. With a pre-production series, and this is its main purpose, statistical tests can be performed. Only then can reliable, quantitative quality statements be made. How many units should be produced and tested? If your company does not employ a quality engineer, it is up to you to suggest a quantity.

The size of a pre-production lot depends on whether it can be sold (i.e., ultimately, on the result of the test itself). If you knew that everything was fine, you could produce a batch that would be sold later, but then you would not have to do the test. The opposite is to assume that you will at least find something critical that needs to be improved. If you ask your boss for a large pre-production lot size, you must consider that there may not be budget left for a second pre-production run if major defects are found. Statistically, a large lot size of about 100 boards would be optimal. Tactically, it is better to choose a lot size that is just statistically acceptable, say 20 to 25 units.

Tests Performed on the Pre-Production Series

The following tests are common for pre-production:

- Statistical timing analysis.

- Signal statistics as a function of temperature.

- Vibration and shock tests.

Traditionally, pre-production testing is performed by the pre-production manufacturer, often the assembler. Therefore, sketch out a test concept early on, and ask for a quote for testing along with a quote for production. The detailed test program does not have to be ready at the time of the request. However, it is best to discuss the details of the tests with the selected manufacturer/assembler.

A bed-of-nails test is a good post-production check, but it is usually insufficient for a thorough evaluation of the electronic quality of the device, since the maximum measurement frequency is about 50 MHz. Instead, for pre-production, you should use high-quality test equipment, such as gigahertz oscilloscopes and spectrum analyzers, for which adapters are made to match the connectors on the board. The

206 Testing and Verification

best way to do this is to think about pre-production testing from the very beginning of the project. Create high-quality test connection schemes that go beyond prototype-style hacks. For example, look for ease of use, such as reverse polarity protection, easy latching and un-latching, and easy one-sided attachment to the flat board. Look for stability against lever forces and contact reliability (i.e. number of mating cycles and optional latching).

Obviously, vibration testing is mandatory if the device is going to be used in a vibrating environment, such as a car. However, shaker tests are generally advisable because they can reveal poorly soldered pins and components that are not mechanically fixed by the pins. Therefore, pre-production boards should be shaken on their own, or better yet, already mounted in a case.

Vibration testing at discrete frequencies is not enough. The resonances and harmonics of the board are critical. It is difficult to determine exactly where they are in advance. A frequency sweep is recommended. For automotive products, the vibration spectrum used is approximately constant from 10 to 1000 Hz and with respect to shock with 1 to 3 g. Phoenix Contact tests its connectors for railway applications with broadband vibration from 5 to 150 Hz and with 0.58 g shock. The shocks are applied in all three spatial directions, each with positive and negative polarity.

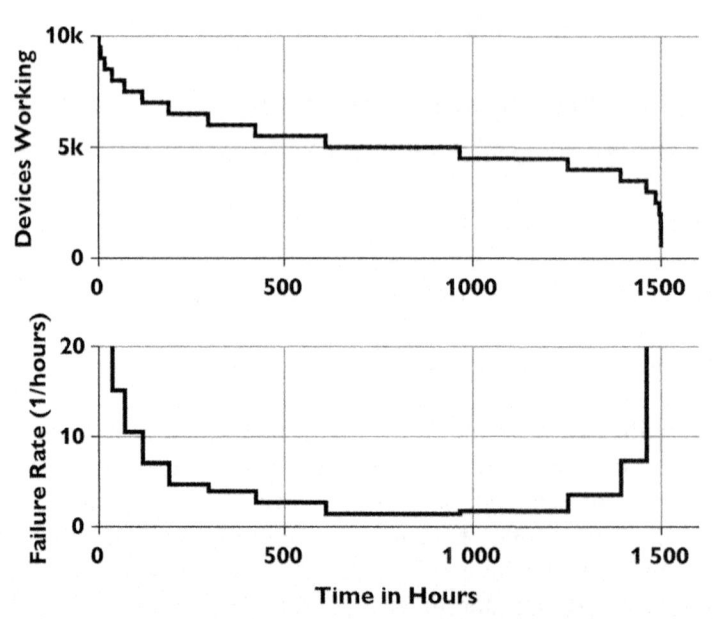

Figure 119. Typical failure rate graph. To determine the failure rate, the same number of consecutive failures must be grouped together. The failure rate is obtained by dividing this number by the time interval covered by the group.

A pre-production lot allows for burn-in testing. During this process, the entire pre-production lot is subjected to extreme temperatures, humidity, and vibration, and is often turned on and off mechanically and electrically to induce early failures. Typically, a predicted failure rate curve can be plotted one week after the start of a burn-in. To understand the implications of the results, let us look at failure statistics for electronic devices.

Failures in technical equipment usually follow the so-called "bathtub curve" (Figure 119): From a production lot, a relatively large number of devices initially fail due to manufacturing defects ("early failures") until a constant failure rate (here: failures per hour) is established. After a certain time, the failure rate increases again due to the aging of the devices.

The term "useful lifetime" is defined using the bathtub curve as the period of constant failure rate in the failure rate diagram. In contrast, the term "lifespan" is not uniformly defined in the literature:

- Sometimes it means the useful life.

- Sometimes it means the time from the first commissioning (i.e., useful life plus period of early failures).

- Sometimes, it is used for the period of cumulative operating times until 0.5% or 1% of the equipment has failed.

- In physics, the characteristic lifetime means the time until 63.2% of radioactive atoms decay. Of course, this is not a reasonable measure of the operating time for a batch of equipment.

Hence, it is better to use the term useful life.

If we assume that the causes of individual failures are statistically independent (e.g., a capacitor fails in one electronic system and a transistor fails in another), then the failure rate could theoretically be determined based on a single PCB. However, we would then have to wait a much longer period of time for a sufficient number of failures to occur for evaluation.

Confusing and different definitions of failure rates are often found in the literature and on the Internet. The definition used here corresponds to the intuitive understanding of a "failure rate" in such a way that a constant failure rate value results if the time between failures is constant. Other definitions of rates that are also confused with failure rates are hazard rate and rate of occurrence of failures (ROCOF).

Since the failure rate is constant over the useful life, we can speak of "the" failure rate of a batch of produced devices, and it can be calculated as follows:

- Total operating hours = batch size × operating hours

208   Testing and Verification

- Number of failing devices = value in ppm × batch size/10^6

- Failure rate = number of failed devices/total operating hours

- An auxiliary unit failure in time (FIT) is often used. The following applies: 1 FIT = $1/(10^9 \text{ h})$.

Finally, the mean time between failures (MTBF) value is the reciprocal of the failure rate.

Keep in mind that MTBF is a purely statistical measure! For example, an MTBF of 20 million hours means that in the operation of a batch of PCBs, one PCB will fail every 20 million cumulative hours of operation. If we assume that 10,000 PCBs run for exactly the same amount of time (e.g. 2,000 hours per year), we will get the first failure after one year. However, an MTBF value for an individual board is meaningless because it does not indicate when a particular board will fail, only that it is likely that a board will fail after the MTBF time. Therefore, MTBF is most useful to the manufacturer in determining warranty periods and stocking spare parts.

Note: If the device is not repaired, the correct term for the value would not be MTBF, but "mean time to failure" (MTTF). In practice, however, MTBF is often referred to simply as MTBF.

It can happen that you are requested to estimate an MTBF/MTTF value for your PCB. There is a simple, conservative approach based on the assumption that the failure of any component causes the failure of the system. In this case, the device failure rate is equal to the sum of the component failure rates. But how do you get component failure rates? The easiest free source for component failure rates is a freeware program from ALD Ltd. called the "ALD MTBF Calculator" [R080]. The program contains failure rates from 26 recognized sources. One of the best-known written sources of failure rates is the U.S. Department of Defense *MIL-HDBK-217 manual* [R002], which is still useful, although it has not been updated since 1995. The manual contains a number of empirically developed failure rate models based on historical component failure rates for a wide range of component types.

From the *MIL-HDBK-217 manual*, the average failure rates shown in Table 34 can be extracted according to "parts stress" at 25°C. From this, you can identify the four most critical components in terms of MTBF/MTTF:

- Relays

- Electrolytic capacitors

- Integrated circuits

- Connectors

Component	FIT = 1/10⁻⁹ h
Relay, low switching number	150
Power MOSFET	50
Electrolytic capacitor	20
Microcontroller	10
Plug contact, infrequent plugging	10
Operational amplifier	5
Solder connection	0.5
Metal film resistor	1
Tantalum chip capacitor	1

Table 34. Common Component Failure Rates.

The high failure rate of integrated circuits is due to the high density of integration and the fact that the failure of a single transistor/memory cell is expected to bring down the system. In reality, integrated circuits have much lower failure rates. However, it is advisable and within the scope of the design to use as few electrolytic capacitors, connectors and relays as possible, or to pay attention to the quality of these components and not to save money on them

Conservative estimation of MTBF/MTTF may result in a value that is considered too low. One way to achieve a higher MTBF is to use FIT values that are as accurate as possible, rather than averages, and to use values that match the device type exactly. For example, a metal film resistor can be distinguished from a carbon resistor. FIT values that more accurately match the device can be much smaller than the average values. For example, for metal film resistors, you will find FIT values in the range of 0.1 instead of the average value of 1 FIT for resistors in general.

In addition, there are significant differences in the data from different databases due to the temperature and other stress conditions at which they were tested. Note that most data are based on measurements at 20°C or 25°C. However, you often have a higher temperature inside the device. This increase in temperature will result in a lower MTBF. How MTBF scales with temperature depends on the part. Such scaling is taken into account in a more complex analysis.

A more complex failure rate calculation usually results in a higher MTBF estimate because it can be less conservative. Specialized software is available for more complex calculations. However, it is questionable whether this is worthwhile. More complex calculation models may come closer to the actual MTBF, but the basic inaccuracy is still very high. For more accurate calculations, it is best to contact technical institutes. It is not recommended to perform a complicated estimation for people who have nothing to do with such estimations.

Instead of trying to increase an MTBF value that is too low, you can try to convince the customer that any estimated MTBF value is questionable. On the other hand, the actual MTBF can certainly be increased as follows:

- Use high quality products for high failure rate components, e.g. electrolytic capacitors with at least 105°C maximum ambient temperature and low ESR.

- Investing in thermal management to reduce the temperature rise; see the corresponding chapter for more information; this measure does not appear in the simple MTBF estimate.

- Introducing redundancy at the functional level, e.g., error detection/correction of data transmission; this action also does not appear in the simple MTBF estimate.

From the above, the following can be concluded: Simple MTBF estimates are usually not very meaningful; complicated ones are expensive with no guarantee of better accuracy. Therefore, try to dissuade the customer from specifying an MTBF number and instead include a section in the specifications that deals with measures to be taken to increase the MTBF. More important than a rough estimate is good design and test practice!

Production Microcontroller Programming

The following options are discussed for programming the microcontroller in production:

- Preprogrammed by the manufacturer. This is common for very high volume, low cost devices. You may not really have a choice here, as it would simply be too time consuming to program the microcontrollers after assembly. There are countless stories of the wrong version being used for programming by the manufacturer, or of a bug being found just days after a large lot has been produced.

- A common solution for mid-size lots is an edge connector. Since this connector is used only once, a common gold finish is sufficient. Hard gold plating is not required. However, the receptacle must be able to withstand many mating cycles. For the latter, an SD memory card socket is a good example.

- The header connector, which I previously advised against using as a programming connector for the microcontroller in the development stage, is a less problematic solution here, where we choose a 90° angled type and place it on an edge. There are no polarization issues here, and if we move the connector as far over the edge as possible, it will have less impact on space consumption.

- Some bed-of-nail testers today allow microcontroller programming, check with your test center. The programming speed can be very slow due to long lines. This is not suitable for large lots.

Production Tests

After a long period of development and many tests on prototypes and maybe even several pilot series, the time has come: release for production! However, it is illusory to hope that all devices produced in the future will work. For economic reasons, production sites will be changed and parts that are no longer available from one supplier will be purchased from another. Yes, even the pick-and-place machine itself misplaces a part from time to time.

Pilot series testing should be repeated periodically in production. Continuous testing of all produced boards, as in the pre-production phase, is usually not economically feasible and is reserved for special applications (e.g. medical devices). If there is an intelligent component on the board, such as a microcontroller or FPGA, self-testing is common. Test signals on the interfaces can be used to comprehensively test the function of the PCB.

It is best to consider production when creating test adapters for the pre-production run. The same adapters may be used in production too. Accordingly, the production adapters must be of even higher quality than the pre-production adapters. Pay special attention to the number of mating cycles. Expensive pre-production measurement devices can be replaced by self-test analysis options. However, this must be prepared for: a useful self-test often requires additional board connections, for example to sample a supply voltage or a signal with the microcontroller's ADC. Without an intelligent device, or for more complete testing, a bed-of-nails test can be performed. This must also be prepared as described in the next section.

Bed-of-nails testing is still the industry standard, but it is limited to about 50 MHz. For example, the QmaxV250-TD6 Automated Test Equipment (ATE) specifies a maximum digital pattern rate of 25 MHz for both the test signal and the PCB signal to be measured. The analog bandwidth of this tester was specified at 48 MHz. The Takaya APT-1400 F flying probe tester specifies a 10 MHz limit for each measurement. Note that the frequency does not indicate the slope. A trace may need to be terminated even at comparatively low frequencies. If this is the case, the use of a bed-of-nails or flying probe tester for this connection is tricky and must be clarified.

The ATE QmaxV250-TD6 mentioned above has a 14-bit ADC, which, if we drop the last bit, results in a resolution of about 400 µV. For voltage level measurements, this resolution is unnecessarily good. For current measurements, this means a resolution of 4 mA on a 0.1 Ω shunt. The Takaya APT-1400 F flying probe tester has a resolution of ±200 nA, but this only applies to DC measurements through the device itself (i.e., with the live connection disconnected). The voltage resolution of the ATP-1400 F is 2 mV, which is equivalent to a current resolution of only 20 mA using a 0.1 Ω shunt. Enabling this type of test requires some schematic and layout preparations that are difficult to implement later. Check with your manufacturer.

The limitations of bed-of-nails and flying probe testing often lead to a dedicated test system. A typical setup consists of a sliding load onto which the PCB can be snapped. The sliding movement connects a PCB edge connector or a 90°-angled header on the PCB to a fixed counterpart. This mating part is mounted on a specially designed test system board. The test system applies the supply voltage and starts the test program. This allows even complex interactions with the DUT to be executed and tested. Finally, a green LED lights up. The operator retracts the board and removes it with one simple movement. This is an elegant and fast test procedure.

Why not test a PCB panel as a whole? Some companies are strongly opposed to this approach. The mechanical stress of separating the PCBs—cutting or breaking them apart—can break some boards that have previously passed all the panel-stage tests. On the other hand, when there are many small PCBs in a panel, panelized testing is very attractive. Figure 120 shows two methods of panel testing:

- Use a via placed in the path of the V-groove to transfer power and signals to each PCB. A V-groove is only 30% deep on either side of the panel, leaving an electrical connection through the copper via wall.

- In the break-away situation, leave a mouse bite to route one trace from the edge of the panel to the PCBs. Do not route two traces on both sides of the board, as this could cause a short circuit in the breakaway situation. As in the example in Figure 120, use two left mouse bites for two traces.

In panel testing, it is especially important to move ceramic capacitors as far away from the edge as possible, or to use ceramic capacitors with "soft pads". Ceramic capacitors have a tendency to crack and short

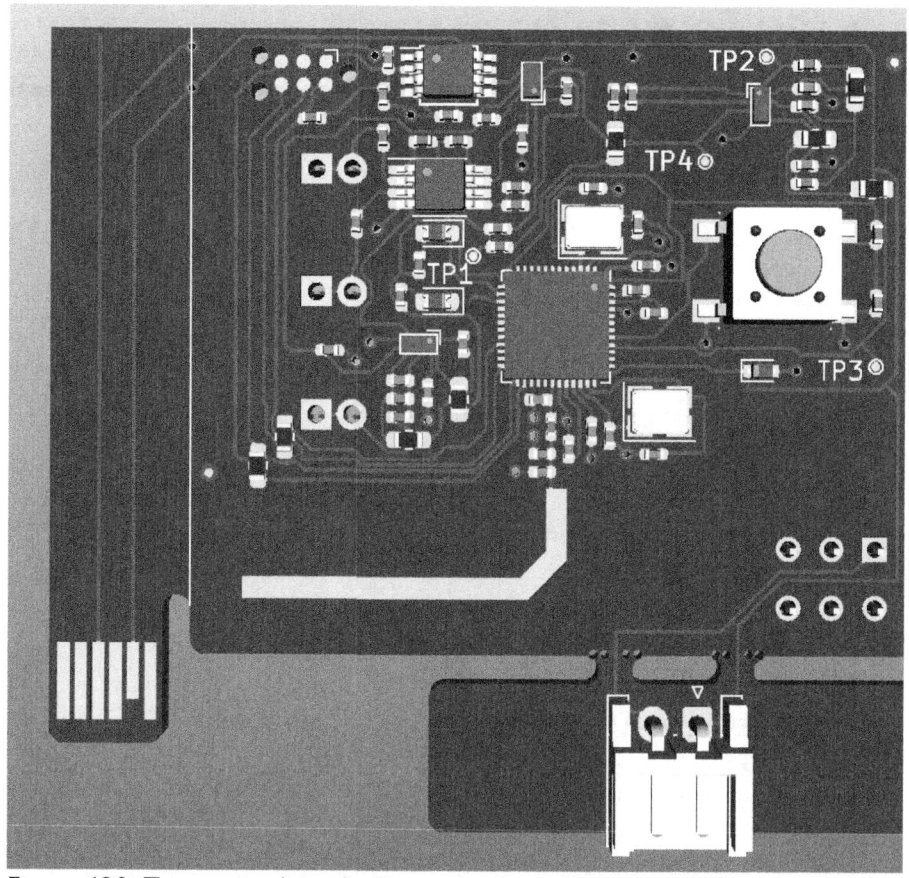

Figure 120. Test entire boards. Power and signal traces can be routed through V-grooves using vias and across break axes without a mouse bite. Not all companies allow full panel testing because there is a mechanical stress step after the final electrical test when the boards are separated.

out under mechanical stress. If you must place a capacitor at the edge, consider adding another capacitor in series, doubling the capacitance values, and placing the second capacitor at a 90° angle to the first.

Burn-in After Production

Post-production burn-in tests are controversial. Because of extreme stress, they shorten the life of the entire batch [R081]. On the other hand, if customers receive too many dead-on-arrival devices, the supplier is considered a low-quality source. Therefore, performing a burn-in test and determining how long to run it is also a matter of company policy, unless the customer specifically requests and defines a burn-in, as NASA does [R001, p. 4]. Note that burn-ins do not change the MTBF, but only eliminate early failures, i.e., the beginning of the curve in Figure 119.

List of Abbreviations

AC—Alternating Current
ADC—Analog-to-Digital Converter
ASIC—Application-Specific Integrated Circuit
AVCC—Analog Voltage Common Collector (Analog Supply Voltage)
BGA—Ball Grid Array
BOM—Bill of Materials
CAN—Controller Area Network
CISPR—Comité International Spécial des Perturbations Radioélectriques (International Special Committee on Radio Interference)
CMOS—Complementary Metal-Oxide-Semiconductor
CPLD—Complex Programmable Logic Device
DAC—Digital-to-Analog Converter
DC—Direct Current
DMA—Direct Memory Access
DRAM—Dynamic Random-Access Memory
DSP—Digital Signal Processor
DSUB—D-Subminiature (a type of electrical connector)
EEPROM—Electrically Erasable Programmable Read-Only Memory
EMC—Electromagnetic Compatibility
EMI—Electromagnetic Interference
ESD—Electrostatic Discharge
ESR—Equivalent Series Resistance
ETSI—European Telecommunications Standards Institute
FPGA—Field-Programmable Gate Array
FRAM—Ferroelectric Random-Access Memory
GND—Ground
GPIO—General Purpose Input/Output
GPS—Global Positioning System
HAL—Hardware Abstraction Layer
HDI—High-Density Interconnect
IBIS—Input/Output Buffer Information Specification
IC—Integrated Circuit
ICE—In-Circuit Emulator
INA—Instrumentation Amplifier
IPC—Institute for Interconnecting and Packaging Electronic Circuits
JEDEC—Joint Electron Device Engineering Council

JTAG—Joint Test Action Group
LCD—Liquid Crystal Display
LDO—Low-Dropout Regulator
LED—Light-Emitting Diode
LVDS—Low-Voltage Differential Signaling
MCU—Microcontroller Unit
MELF—Metal Electrode Leadless Face
MEMS—Microelectromechanical Systems
MISO—Master In Slave Out
MLCC—Multilayer Ceramic Capacitor
MOSFET—Metal-Oxide-Semiconductor Field-Effect Transistor
MTBF—Mean Time Between Failures
MTTF—Mean Time To Failure
OLED—Organic Light-Emitting Diode
PCB—Printed Circuit Board
PCMCIA—Personal Computer Memory Card International Association
PMOS—P-Channel Metal-Oxide-Semiconductor
PTC—Positive Temperature Coefficient (or a device with a PTC)
PWM—Pulse Width Modulation
QFN—Quad Flat No-Leads
RAM—Random Access Memory
RED—Radiated Emission Directive
RISC—Reduced Instruction Set Computing
RoHS—Restriction of Hazardous Substances
SAR—Successive Approximation Register
SDRAM—Synchronous Dynamic Random-Access Memory
SIMM—Single Inline Memory Module
SMD—Surface Mount Device
SPI—Serial Peripheral Interface
SPICE—Simulation Program with Integrated Circuit Emphasis
SPST—Single Pole Single Throw
THT—Through-Hole Technology
TTL—Transistor-Transistor Logic
UART—Universal Asynchronous Receiver-Transmitter
USB—Universal Serial Bus
VBUS—Voltage Bus (in USB context)
VCC—Voltage at the Common Collector (used for power supply voltage)
VDD—Voltage Drain to Drain (used in MOSFETs and other semiconductor devices)

Bibliography

R001   Dr. Kusum Sahu, "EEE-INST-002: Instructions for EEE Parts Selection, Screening, Qualification, and Derating," Goddard Space Flight Center, Greenbelt, MD, April 2008.

R002   Department of Defense, MIL-HDBK-217F (Reliability Prediction of Electronic Equipment), Washington, DC: Office of the Secretary of De-fense, 1991.

R003   Pentair, Spec-00488 C.

R004   J. Adam, "New Correlations Between Electrical Current and Temperature Rise in PCB Traces," 20th IEEE SEMI-THERM Symposium, 2004.

R009   John R. Barnes, Robust Electronic Design Reference Book, Volume 1, Dordrecht, Netherlands: Kluwer Academic Publishers, March 30, 2004.

R012   National Semiconductor, Application Note 1229.

R013   MIL-STD 275E, Dec. 1984.

R014   IPC-2221A, May 2003.

R015   International Rectifier, Application Note AN-994.

R016   S.F. Shuler et al, "Design of plastic enclosures for thermal management," ANTEC '97, 1997, pages 3235–3238.

R017   Fairchild Semiconductor, Application Note AN1028.

R018   Wakefield Engineering, Introduction to Thermal Management, Technical Discussion.

R019   "Heat Sink Testing Methods and Common Oversights," Qpedia, Issue II, Number XII, Pages 18–22, Jan. 2009.

R023   F. Alferink, Temperature Measurements, undated. [Online]. Link: https://meettechniek.info/measuring/temperature.html. [Accessed Dec. 2018]

R025   Catalog of requirements 3, Requirements for machine emissions, Federal Institute for Occupational Safety and Health, undated.

R026   J. C. Coyne, "An Approximate Thermal Model for Outdoor Electronics Cabinets," The Bell System Technical Journal, issue 61, number 2, pages 227–246, February 1982.

R027   Thermal Resistance Theory and Practice, Munich, Germany: Infineon Technologies AG, 2000.

R028   Linear Technology, Thermal Resistance Table.

R029   M.T. Zhang, M.M. Jovanovic, F.C. Lee, "Design and analysis of thermal management for high-power-density converters in sealed enclo-sures," APEC '97, 1997, pages 405–412.

R031   On Semiconductor, "Human Body Model vs. IEC 61000-4-2."

R032   K. Loukil and K. Siala, "EMC standards," ITU Training on Conformance and Interoperability for ARB Region, 2013.

R034   Silicon Labs, Application Note AN376.

R035   On Semiconductor, Application Note AND8230/D.

R036   On Semiconductor, Application Note AND8232/D.

R037   Silicon Image, Design Guide for the Control of ESD in the eSATA Interface.

R038   J. Dunnihoo, ESD Protection for High-Speed I/O Signals, 2010. [Online]. Link: http://www.ce-mag.com/archive/03/ARG/dunnihoo.html. [Accessed Dec. 2010.]

R039   L. Robinson, ESD protection tips to improve reliability, 2006. [Online]. Link: http://eetasia.com. [Accessed Dec. 2006.]

R040   Cypress Semiconductor Corp, Electro-static Discharge (ESD) Tutorial.

R041   Elmac Services, "Some considerations for ESD testing".

R042   O. Holenstein, "Teseq—EMC Symposium Zurich 2010," 2010.

R044   Ch. Marak, J. Colby, Four Ways to Enhance ESD Protectio After Your Design Flunks Its ESD Test, 2015. [Online]. Link: http://machinedesign.com/technologies/four-ways-enhance-esd-protect. [Accessed July 2015.]

R045   Ohmite, data sheet OD/OF/OA Series.

R046   D. C. Smith, Effect of High Voltage Pulses on Resistors—ESD and EFT, 2012. [Online]. Link: http://emcesd.com. [Accessed Oct. 2012.]

R048   Maxim Integrated, "Maxim Leads the Way in ESD Protection."

R049   C. Rostamzadeh, F. Canavero, F. Kashefi and M. Darbandi, Effectiveness of multilayer ceramic capacitors for electrostatic discharge protection, 2012. [Online]. Link: https://incompliancemag.com/article/effectiveness-of-multilayer-ceramic-capacitors-for-electrostatic-discharge-protection/. [Accessed May 2012.]

R050   J. Scanlon and K. Rutgers, Safeguard Your RS-485 Communication Networks from Harmful EMC Events, 2013. [Online]. Link: https://www.analog.com/en/analog-dialogue.html. [Accessed May 2013.]

R051   ST Microelectronics, Application Note AN4275.

R052   ATEC Corp, IEC 61000-4-5: Testing and Measurement techniques—Surge Immunity Test, undated. [Online]. Link: http://www.atecorp.com/compliance-standards/iec-standards/iec-61000-4-5-testing-andmeasurement-techniques-s.aspx. [Accessed in May 2012.]

R053   Semtech International AG, Application Note AN96-07.

R056   D. A. Weston, Electromagnetic Compatibility Principles and Applications, New York, NY: Marcel Dekker, Inc, 2001.

R057   N. Ellis, Electrical Interference Handbook, 2nd edition, Oxford, MA: Newnes, 1998.

R058   T. Williams, The Circuit Designer's Companion, 2nd edition, Oxford, MA: Newnes, 2005.

R059   M. I. Montrose, Printed Circuit Board Design Techniques for EMC Compliance, Hoboken, NJ: Wiley, 2000.

R066   Recom, Application Notes 2009.

R067   Circuit Design Know It All, Oxford, MA: Newnes, 2008.

R068   D. Brooks and D. Graves, "Current Carrying Capacity of Vias," UltraCAD Design, Inc, Issaquah, WA, Jan. 2003.

R069   PCB Via Calculator, 2006. [Online]. Link: http://circuitcalculator.com/wordpress/2006/03/12/pcb-via-calculator/. [Accessed July 2015.]

R070   Texas Instruments, "Constructing Your Power Supply-Layout Considerations."

R071   Ch. Kitchin, Demystifying single-supply op-amp design, 2002. [Online]. Link: www.ednmag.com. [Accessed July 2015]

R072   W. J. Dally, EE273 Lecture 18: "On-Chip Power Distribution," Stanford Univ., Stanford, CA, Nov. 25, 1998.

R073   Texas Instruments, Application Note SLOA091.

R076   Jack G. Ganssle, "A Guide to Debouncing," Ganssle Group, Baltimore, MD, June 2008.

R078   B. Baker, A Baker's Dozen: Real Analog Solutions for Digital Designers, Oxford, MA: Newnes, 2005.

R079   Texas Instruments, Application Note SCBA004C.

R080   ALD Ltd, Free MTBF Calculator, 2015. [Online]. Link: http://aldservice.com/en/Reliability-Software/free-mtbf-calculator.html. [Accessed July 2015.]

R081   How Long Should You Burn In a System?, 2006. [Online]. Available: https://www.weibull.com/hotwire/issue69/relbasics69.htm. [Accessed July, 2015]

R084   B. Pease, Troubleshooting Analog Circuits, Burlington, MA: Newnes, 1993.

R086   Vishay Sfernice, data sheet CHP, HCHP.

R087   Vishay Intertechnology Inc., Appl. Note Frequency Response of Thin Film Chip Resistors.

R089   E. Bogatin, Signal bandwidth from clock frequency: rule of thumb 2, 2013. [Online]. Link: https://www.edn.com/rule-of-thumb-2-signal-bandwidth-from-clock-frequency/. [Accessed May 2015.]

R095   Tzong-Lin Wu, Spectra of Digital Waveform, National Taiwan University, Taipei City.

R096   H.Kapitza, Some Remarks on Shielding, DESY, FLA, Hamburg, Germany, Oct. 9, 2006.

R097   Würth Elektronik, "The Protection of USB 2.0 Applications".

R098   Murata Manufacturing Co, Ltd, "Noise Suppression by EMIFIL®".

R105   D. Brooks, Signal Integrity Issues and Printed Circuit Board Design, Upper Saddle River, NJ: Prentice Hall, 2008.

R110   Fan-out, 2015. [Online]. Link: https://en.wikipedia.org/wiki/Fan-out. [Accessed May 2015.]

R111   Texas Instruments, Application Note SCYB004B.

R118   JESD8C.01, Sept. 2007.

R121   L.Ritchey, RIGHT THE FIRST TIME—Vol. I, Bodega Bay, CA: Speeding Edge, 2008.

R125   F. Hillebrand, High-Speed Seminar, ZHAW, Winterthur, Switzerland, May 28, 2003.

R126   L. Ritchey, SIGNAL INTEGRITY AND HIGH SPEED SYSTEM DESIGN, Bitburg, Germany, May 2009.

R127   Analog Devices, Application Note AN-397.

R136   B. Baker, "The IBIS model: A conduit into signalintegrity analysis, Part 1," Texas Instruments Inc., Dallas, TX, June, 2010.

R137   B. Baker, "The IBIS model: A conduit into signalintegrity analysis, Part 2," Texas Instruments Inc., Dallas, TX, June, 2010.

R138   B. Baker, "The IBIS model: A conduit into signalintegrity analysis, Part 3," Texas Instruments Inc., Dallas, TX, June, 2010.

R139   Fairchild Semiconductor, Application Note AN-393.

R141   Lee W. Ritchey, Differential Signaling Doesn't Require Differential Impedance, 2005. [Online]. Available: http://www.speedingedge.com/PDF-Files/diffsig.pdf. [Accessed July, 2015]

R142   Murata Manufacturing Co., Ltd., Appl. Note SMD/BLOCK Type EMI Suppression Filters EMIFIL.

R144   Maxim Integrated, Application Note 4266.

R145   W. W. Coblentz, C. W. Hughes, "Emissive Tests of Paints for Decreasing or Increasing Heat Radiation From Surfaces," Washington, DC, March 1924.

R146   G. F. Hundy, A. R. Trott, T. C. Welch, *Refrigeration and Air-Conditioning*, Oxford, UK: Butterworth-Heinemann, 2008.

R147   Xiaodong He, Yibin Li, Lidong Wang, Yue Sun, Sam Zhang, "High emissivity coatings for high temperature application: progress and prospect," Thin Solid Films, issue 517, pages 5120–2129, 2009.

R148   EPA-450/3-85-019a, Surface Coating of Plastic Parts For Business Machines-Background Information for Proposed Standards, U.S. Environmental Protection Agency, December 1985.

R149   R. H. Hoffman, "Spaceflight Performance of Silver Coated FEP Teflon," Greenbelt, MD, April 1973.

R150   Thermal Resistance Theory and Practice, Munich, Germany: Infineon Technologies AG, 2000.

R156   Maxim Integrated, Application Note 3500.

R157   J. Ardizzoni, ESD Diode Doubles as Temperature Sensor, AnalogDialog, Vol. 41, Nov, 2007. [Online]. Link: https://www.analog.com/en/analog-dialogue. [Accessed Dec. 2018.]

R158   Thermal Engineering Associates, Application Note TB-02.

R159   Texas Instruments, Application Note SBOA277.

R160   Texas Instruments, Application Note SPRA953C.

R161   On Semiconductor, Application Note AN1040/D.

R162   International Rectifier, Application Note AN-1012.

R163   Fairchild Semiconductor, Application Note AB-9.

R164   "Heat Sink Design Facts & Guidelines for Thermal Analysis," Wakefield-Vette, Pelham, NH, Nov. 25, 1998.

R165   JESD51-2A, Jan. 2008.

R166 P. Rako, Hot, cold, and broken: thermal-design techniques, March 29, 2007. [Online]. Link: https://www.edn.com/design/analog/4316239/Hot-cold-and-broken-Thermal-design-techniques. [Accessed Dec. 2018.]

R167 J. Titus, Design electronics for cold environments, Dec. 17, 2012. [Online]. Link: https://www.ecnmag.com/article/2012/12/design-electronics-cold-environments. [Accessed Dec. 2018.]

R168 A. Koss, How heat sink fin spacing plays a key role in natural convection cooling and how you can optimize it, Sept. 8, 2010. [Online]. Link: https://www.qats.com/cms/2010/09/08/how-heat-sink-fin-spacing-plays-a-key-role-in-natural-convection-cooling-and-how-you-can-optimize-it/. [Accessed Dec. 2018.]

R171 Programming devices at In-Circuit Test? 2011. [Online]. Link: , https://blog.asset-intertech.com/test_data_out/2011/11/programming-devices-at-in-circuit-test.html. [Accessed June 2019]

R181 Bonnie Baker and John Z. Wu, Overcoming the challenges of linking A/D converters and microcontrollers via long transmission lines, 2008. [Online]. Availa-ble: https://www.eetimes.com/overcoming-the-challenges-of-linking-a-d-converters-and-microcontrollers-via-long-transmission-lines-part-1-of-2/. [Ac-cessed July, 2015]

R182 Resistor Sensitivity to Electrostatic Discharge (ESD), VISHAY INTERTECHNOLOGY, INC., Malvern, PA, Sept., 2011.

R185 National Semiconductor, Application Note AN-779.

R188 Brian C. Wadell, Transmission Line Design Handbook , Norwood, MA: Artech House Inc., 1991

R190 D. Brooks, "90 Degree Corners: The Final Turn," Printed Circuit Design Magazine, January 1998.

R191 H. W. Ott, *Electromagnetic Compatibility Engineering*, Hoboken, New Jersey: Wiley, 2009.

R192 KEMET, Application Notes for Tantalum Capacitors.

R193 D. Meeker, Finite Element Method Magnetics. [Online]. Link: https://www.femm.info [accessed April 2021]

R194 L. Li, "Computation of power plane pair inductance, measurement of multiple switching current components and switching current measurement for multiple ICs with an island structure" (2012). Masters Thesis, Missouri University of Science and Technology.

R195 K. Carver and J. Mink, "Microstrip antenna technology," in IEEE Transactions on Antennas and Propagation, vol. 29, no. 1, pp. 2-24, January 1981.

R197 J.E. Arsenault, John Alvin Roberts, *Reliability & Maintainability of Electronic Systems*, Potomac, Maryland: Computer Science Press, 1980.

R199 A. Zeverev, *Handbook of Filter Synthesis*, Hoboken, New Jersey: John Wiley & Sons 2005.

R200 A. B. Williams and F. J.Taylor, *Electronic Filter Design Handbook*, McGraw-Hill, 2006.

R201 Murata, Precautions for measuring the capacitance of chip multilayer ceramic capacitors, undated. [Online]. Available: https://www.murata.com/en-global/products/capacitor/ceramiccapacitor/library/solution/insufficient. [Ac-cessed December, 2023]

The missing reference numbers are not used in this book, but were in an older edition. For the sake of comparison, reference numbers have not been renumbered for this edition.

Printed in Great Britain
by Amazon